NATIONAL
ACADEMIES
Sciences
Engineering
Medicine

NATIONAL
ACADEMIES
PRESS
Washington, DC

An Assessment of Native Seed Needs and the Capacity for Their Supply

Final Report

Committee on an Assessment of Native Seed
Needs and Capacities

Board on Agriculture and Natural Resources

Division on Earth and Life Studies

Committee on National Statistics

Division of Behavioral and Social Sciences
and Education

A Consensus Study Report

NATIONAL ACADEMIES PRESS 500 Fifth Street, NW Washington, DC 20001

This activity was supported by Contract No. 140L0618C0052 between the National Academy of Sciences and the Department of the Interior's Bureau of Land Management. Any opinions, findings, conclusions, or recommendations expressed in this publication do not necessarily reflect the views of any organization or agency that provided support for the project.

International Standard Book Number-13: 978-0-309-69025-6
International Standard Book Number-10: 0-309-69025-0
Digital Object Identifier: https://doi.org/10.17226/26618
Library of Congress Control Number: 2023934994

Cover photo: Lemmon's marigold (*Tagetes lemmonii*), BLM AZ930, Seeds of Success.

This publication is available from the National Academies Press, 500 Fifth Street, NW, Keck 360, Washington, DC 20001; (800) 624-6242 or (202) 334-3313; http://www.nap.edu.

Suggested citation: National Academies of Sciences, Engineering, and Medicine. 2023. *An Assessment of Native Seed Needs and the Capacity for Their Supply: Final Report.* Washington, DC: The National Academies Press. https://doi.org/10.17226/26618.

The **National Academy of Sciences** was established in 1863 by an Act of Congress, signed by President Lincoln, as a private, nongovernmental institution to advise the nation on issues related to science and technology. Members are elected by their peers for outstanding contributions to research. Dr. Marcia McNutt is president.

The **National Academy of Engineering** was established in 1964 under the charter of the National Academy of Sciences to bring the practices of engineering to advising the nation. Members are elected by their peers for extraordinary contributions to engineering. Dr. John L. Anderson is president.

The **National Academy of Medicine** (formerly the Institute of Medicine) was established in 1970 under the charter of the National Academy of Sciences to advise the nation on medical and health issues. Members are elected by their peers for distinguished contributions to medicine and health. Dr. Victor J. Dzau is president.

The three Academies work together as the **National Academies of Sciences, Engineering, and Medicine** to provide independent, objective analysis and advice to the nation and conduct other activities to solve complex problems and inform public policy decisions. The National Academies also encourage education and research, recognize outstanding contributions to knowledge, and increase public understanding in matters of science, engineering, and medicine.

Learn more about the National Academies of Sciences, Engineering, and Medicine at **www.nationalacademies.org**.

COMMITTEE ON AN ASSESSMENT OF NATIVE SEED NEEDS AND CAPACITIES

SUSAN P. HARRISON (*Chair*), University of California, Davis
DELANE ATCITTY, Indian Nations Conservation Alliance, El Prado, NM
ROB FIEGENER, Independent Consultant, Corvallis, OR
RACHAEL GOODHUE, University of California, Davis
KAYRI HAVENS, Chicago Botanic Garden, Glencoe, IL
CAROL C. HOUSE, Independent Consultant, Lyme, CT
RICHARD C. JOHNSON, Washington State University, Pullman
ELIZABETH LEGER, University of Nevada, Reno
VIRGINIA LESSER, Oregon State University, Corvallis
JEAN OPSOMER, Westat, Rockville, MD
NANCY SHAW, US Forest Service, Boise, ID (Emeritus)
DOUGLAS E. SOLTIS, Florida Museum of Natural History, Gainesville
SCOTT M. SWINTON, Michigan State University, East Lansing
EDWARD TOTH, Mid-Atlantic Regional Seed Bank, Cortland, NY
STANFORD A. YOUNG, Utah State University, Logan (Emeritus)

Study Staff

ROBIN SCHOEN, Director, Board on Agriculture and Natural Resources
KRISZTINA MARTON, Senior Program Officer
JENNA BRISCOE, Research Associate (until September 2021)
SARAH KWON, Senior Program Assistant (until May 2022)
PAIGE JACOBS, Program Assistant (until November 2022)
SAMANTHA SISANACHANDENG, Program Assistant

Acknowledgments

This Consensus Study Report was reviewed in draft form by individuals chosen for their diverse perspectives and technical expertise. The purpose of this independent review is to provide candid and critical comments that will assist the National Academies of Sciences, Engineering, and Medicine in making each published report as sound as possible and to ensure that it meets the institutional standards for quality, objectivity, evidence, and responsiveness to the study charge. The review comments and draft manuscript remain confidential to protect the integrity of the deliberative process.

We thank the following individuals for their review of this report:

AMY W. ANDO, University of Illinois, Urbana-Champaign
MEGHAN BROWN, Nevada Department of Agriculture
BRIAN M. IRISH, US Department of Agriculture, Agricultural Research Service
KRISTEN OLSON, University of Nebraska, Lincoln
STEVE PARR, Upper Colorado Environmental Plant Center
S. LYNNE STOKES, Southern Methodist University
G. DAVID TILLMAN, University of Minnesota
RUSSELL E. TRONSTAD, University of Arizona
ANDREA WILLIAMS, California Native Plant Society

Although the reviewers listed above provided many constructive comments and suggestions, they were not asked to endorse the conclusions or recommendations of this report, nor did they see the final draft before its release. The review of this report was overseen by **PETER H. RAVEN (NAS)**, Missouri Botanical Garden (Emeritus), and **MAY R. BERENBAUM (NAS)**, University of Illinois at Urbana-Champaign. They were responsible for making certain that an independent examination of this report was carried out in accordance with the standards of the National Academies and that all review comments were carefully considered. Responsibility for the final content rests entirely with the authoring committee and the National Academies.

Preface

Since the publication of our Interim Report in late 2020, the need to strengthen the nation's supply of native seeds for ecological restoration and related purposes has only become clearer. The year 2021 came in just behind 2020 in terms of number of multi-billion-dollar climatic disasters (20 versus 22) and third in total costs (behind 2017 and 2005), with a price tag of $145 billion (www.ncei.noaa.gov/access/billions/). Major climate-related events in 2021 alone included a severe cold wave in the South, massive wildfires and continued drought in the West, flooding in California and Louisiana, three tornado outbreaks, four tropical cyclones, and eight other severe weather events. The increasing magnitude and frequency of such climatic mega-disturbances is straining not only our economy but the recovery capacity of ecosystems, in synergy with other unceasing stresses including invasive species, energy and mineral extraction, urbanization, and land conversion. As the vulnerabilities of humans, wildlife, and critical ecosystem services to these disruptions grow, the need for ecological restoration in the 21st century will continue its trajectory toward a previously unmatched scale. In the United States just as elsewhere in the world, a limited supply of native seeds and other native plant materials is a widely acknowledged barrier to fulfilling our most critical restoration needs.

In our efforts to analyze the nation's system of producing and using native plant materials for restoration and to identify the most impactful steps toward improving the supply, we were not helped by the COVID-19 pandemic. Our meetings, presentations, and information gathering were slowed significantly, and the availability of our committee members, National Academies of Sciences, Engineering, and Medicine staff, and expert informants across the nation were drastically altered by the many changes to people's professional and personal lives. "Nevertheless, we persisted." We are now honored to release what we believe is a well-supported set of key recommendations for improving the native seed supply, backed by findings and conclusions reached through collecting input from native seed producers and users in the public, private, nonprofit, and academic sectors across the United States.

I'd like to thank the committee members who have worked so hard to bring this report to its fruition, along with the staff from the National Academies of Sciences, Engineering, and Medicine. And together, all of us thank the expert informants whose professional dedication to the supply and use of native plant materials made this report possible.

Sincerely,

Susan P. Harrison, *Chair*
Committee on an Assessment of
Native Seed Needs and Capacities

Contents

APPENDIXES

Boxes, Figures, and Tables

BOXES

FIGURES

TABLES

Acronyms and Abbreviations

AOSA	Association of Official Seed Analysts
AOSCA	Association of Official Seed Certifying Agencies
ARS	Agricultural Research Service
BLM	Bureau of Land Management
DOD	Department of Defense
DOI	Department of the Interior
DOT	Department of Transportation
EPA	Environmental Protection Agency
G_0, Gx	Generation zero, Generation x
IDIQ	Indefinite Delivery Indefinite Quantity
NASEM	National Academies of Sciences, Engineering, and Medicine
NIFA	National Institute of Food and Agriculture
NPS	National Park Service
NRCS	Natural Resources Conservation Service
PCRP	Plant Conservation and Restoration Program
PLS	Pure live seed
PMC	Plant Materials Center
REPLANT	Repairing Existing Public Land by Adding Necessary Trees
RFP	Request for Proposal

SCST   Society of Commercial Seed Technologists
SESRC   Social and Economic Sciences Research Center
SI   Source-Identified

USDA   US Department of Agriculture
USFS   US Forest Service
USFWS   US Fish and Wildlife Service
USGS   US Geological Survey

Glossary

Accession: A distinct, uniquely identified sample of seeds or plants.

Adaptive management: A structured process of using management as an experiment, so that new information is gained that reduces uncertainty about the managed system and enables management to improve over time.

Agronomically: Related to growing a crop, with a focus on managing the soil, nutrients, and the physical and biological environment to support crop production.

Cultivar: A named variety of a plant species with distinct genetically based morphological, physiological, cytological, or chemical characteristics, produced and maintained by cultivation.

Ecological restoration: The process of assisting the recovery of an ecosystem that has been degraded, damaged, or destroyed.

Ecoregion: A relatively large unit of land or water that is characterized by a distinctive climate, ecological features, and plant and animal communities.

Ecosystem: A biological community of interacting organisms and their physical environment.

Ecotype: A genetically distinct subset of a species (a population, subspecies, or race) that is adapted to local environmental conditions.

Empirical seed zone: Area within which plant materials are believed to be transferrable with little risk of being poorly adapted to their new location, developed by combining species-specific genetic information on local adaptation with environmental information.

Establishment: The stage at which the seedling has exhausted the food reserves stored in the seed and must grow, develop, and persist independently.

Extractory: Facility for the cleaning, conditioning, and short-term storage of wild-collected seed.

Forbs: Vascular plants that are not woody and also not grasses or members of the grass family, sometimes colloquially called "wildflowers."

Genetically adapted: See Local Adaptation.

Genetically appropriate: Native plant materials that are likely to establish, persist, and promote ecological relationships at a restoration site. Such plants would be sufficiently genetically diverse to respond to changing environmental conditions; unlikely to cause genetic contamination of resident native species; unlikely to become invasive and displace other native species; unlikely to be a source of non-native pathogens; and likely to maintain relationships with other native species.

Germplasm: Living genetic resources such as seeds or tissues that are maintained for breeding, research, and conservation efforts.

Habitat: The place or environment where a plant or animal naturally or normally lives and grows.

Invasive species: A species that is non-native to the ecosystem under consideration and which is likely to cause economic or environmental harm.

Local adaptation: Evolution of genetically distinct traits that make certain populations of a species better able to establish and persist in their particular local environment than other populations of the same species from other locations.

Maladaptation: Having traits that are poorly suited or adapted to a particular situation or set of conditions.

Native plant communities: Recurring assemblages of native plant species associated with particular regions and environmental conditions.

Native plants: Species that occur naturally in a particular region, ecosystem, or habitat, having either evolved there or dispersed there unaided by humans.

Non-native species: Species that have been accidentally or deliberately introduced by humans to a continent, region, ecosystem, or habitat in which they did not previously occur.

Plant materials: Any portion of a plant that can be propagated, including seeds, cuttings, and entire plants.

Provenance: The geographic origin of a seed source.

Provisional seed zone: Area within which plant materials are believed to be transferrable with little risk of being poorly adapted to their new location, developed using climatic and other environmental data, but not using species-specific information.

Pure live seed: The germinable seed in a seedlot. As displayed on the label of a seed bag, it is the weight of the bag minus weeds, impurities, and inviable seed.

Rehabilitation: Restoring a particular function such as erosion control to a damaged or degraded area, using native or non-native species.

Restorative continuum: An array of activities that reduce degradation and support partial to full ecosystem recovery; for example, revegetation, rehabilitation, and ecological restoration.

Revegetation: Restoring plant cover to a damaged or degraded area, using native or non-native species.

Seed bank: A storage facility intended to preserve seeds for the future, which requires low humidity and low temperatures.

Seed certification: A legally controlled system of quality control over seed multiplication and production.

Seed increase: Cultivation of a plant with the goal of obtaining a larger quantity of seeds for future cultivation.

Seed quality: The combination of correct genetic identity, germination fraction, and vigor in a batch of seeds.

Seed viability: The capacity of a seed or batch of seeds to germinate under suitable conditions, including dormant seeds for which dormancy must be broken before viability can be measured by germination.

Seed zone: A mapped area within which plant materials are believed to be transferrable with little risk of being poorly adapted to their new location.

Taxon: A collection of one or more populations of organisms. Taxa are the hierarchical classifications of a species (e.g., species, subspecies).

Traditional Ecological Knowledge: Also called Indigenous Knowledge or Native Science, the evolving knowledge of a specific location acquired by indigenous and local peoples, including relationships between plants, animals, the physical environment, and their uses for activities including but not limited to hunting, fishing, trapping, agriculture, and forestry.

Workhorse species: Native species with the potential for broad use in restoration across a region, selected because they are abundant across a wide range of ecological settings, establish quickly, and support important ecological processes.

Summary

BACKGROUND

Millions of acres of public and private land in the United States are at risk of losing the native plant communities that are central to the integrity of ecosystems. As plant communities decline, the biodiversity they embody is also being rapidly lost, along with a wealth of ecosystem goods and services important to society. There is a long and growing list of threats to plant communities—invasive species, overgrazing, climate change, and altered fire regimes, to name a few—that have collectively accelerated the deterioration of natural landscapes across the country.

Ecological restoration is the process of bringing back native biological diversity and ecosystem function to deteriorated landscapes, one aspect of which involves planting seeds of native plants on the degraded site. This report examines the prospects for developing the large and sustainable supply of native seeds (as a shorthand term for all forms of native propagative plant material[1]) that is needed for many uses but primarily to carry out successful ecological restoration across our nation's landscapes.

More than 20 years ago, as severe wildfires began to burn with increasing frequency and severity in the western United States, Congress urged the Department of the Interior (DOI) to move beyond sowing non-native grass seed for emergency soil stabilization and toward the rapid introduction of native plant species to rebuild natural communities and prevent invasive plant encroachment in newly burned areas. Congress requested the Secretaries of the Interior and Agriculture to prepare "specific plans and recommendations to supply native plant materials for emergency stabilization and longer-term rehabilitation."[2]

In early 2002, the agencies responded with a 17-page plan and five action items:

1. Undertake a comprehensive assessment of needs for native plant materials.
2. Make a long-term commitment to native plant materials production, research and development, education, and technology transfer.
3. Expand efforts to increase the availability of native plant materials.

---

[1] The statement of task (Box S-1) specifically refers to "native plant seed." The report uses native plant seed to encompass not only seeds but other native plant materials, such as containerized stock, bare root seedlings, cuttings, rhizomes, tissue culture, callus material, and other plant propagules.

[2] House Committee on Appropriations Report, to accompany H.R. 2217, Department of the Interior and Related Agencies Appropriations Bill, 2002, 107th Congress.

4. Invest in partnerships with state and local agencies and the private sector.
5. Ensure adequate monitoring of restoration and rehabilitation efforts.

The plan cited the histories of success by US Forest Service (USFS) nurseries in the 1920s to produce conifer tree seedlings and by the US Department of Agriculture's (USDA) Plant Materials Centers in the 1930s Dust Bowl era to develop plant species for soil conservation.

The USFS, Bureau of Land Management (BLM), National Park Service (NPS), Department of Defense (DOD), and US Fish and Wildlife Service (USFWS), many states, and tribal nations have access to native plant communities on lands they manage, which provide the source material for the native seed supply. Seeds collected from these communities can either be used directly or increased through agricultural cultivation before being used as seeds or plants. A key consideration in restoration is for native plants to be genetically adapted to the specific sites or regions where they are used, so recording source locations and maintaining the genetic identity of wild-collected native seeds is critical.

A relatively small segment of the nation's commercial seed industry produces seeds of native plant species for ecological restoration, in an enterprise that demands considerable specialized knowledge and equipment, and that requires several years of lead time to produce a specified batch of seeds. Public agencies and other users of native seed acquire most of their supply from these private growers, using requests for bids, production agreements, and off-the-shelf purchases. Currently, however, users describe the supply of native seeds on the market as severely insufficient. Suppliers describe the buyers' unpredictable levels of demand and excessively short planning horizons as significant obstacles to being able to meet their needs.

ASSESSMENT OF NATIVE SEED NEEDS AND CAPACITIES

This report was requested by the BLM and prepared by a committee of experts appointed by the National Academies of Sciences, Engineering, and Medicine. It was not commissioned as a review of the original 2002 plan, but as an independent assessment of the federal, state, tribal, and private-sector needs and capacity for supplying native plant seeds for ecological restoration and other purposes (see the abbreviated statement of task in Box S-1). This report reflects the product of phase two of the assessment, in follow-up to the committee's interim report released in October 2020.[3] That report presented an overview of the native seed supply chain,

---

BOX S-1
Statement of Task

An ad hoc study committee appointed by the National Academies of Sciences, Engineering, and Medicine will assess federal, state, tribal, and private-sector needs and capacity for supplying native plant seeds for ecological restoration and other purposes. The assessment will focus on the western continental United States and incorporate information from assessments of other US regions, as available, toward the goal of a nationwide perspective. The assessment will be carried out in two phases. In phase one, the committee will conduct fact finding, develop a framework for information gathering for the assessment, and prepare an interim report describing the framework and implementation strategy. In the second phase, the committee will oversee the data- and information-gathering process, analyze the information obtained, and prepare a final report summarizing the committee's findings and conclusions. The final report also will provide recommendations for improving the reliability, predictability, and performance of the native seed supply.

---

[3] An Assessment of Native Seed Needs and the Capacity for Their Supply: Interim Report. The National Academies Press. https://nap.national academies.org/catalog/25859/an-assessment-of-the-need-for-native-seeds-and-the-capacity-for-their-supply.

preliminary observations of some of the challenges facing native seed use and supply, and a strategy for additional information gathering.

In phase two, the committee used the preliminary observations as the basis for two nationwide surveys, the first, of native seed and plant suppliers, and the second, of personnel from state government departments of natural resources, parks, wildlife, transportation, and agriculture.

Activities carried out in 2021 and 2022 also included semi-structured interviews with federal agency personnel, examination of agency records, presentations by participants in the seed supply chain, and reviews of the relevant scientific literature.

THE NATIVE SEED SUPPLY

Federal, State, and Tribal Needs and Capacities

Native seeds are used in pursuit of a wide range of objectives on public land. Table S-1 lists the primary uses of native plant seeds by federal and state agencies. Tribal native seed needs often parallel those of federal and state agencies. The tribes' needs for native plants also include food, spiritual, and medicinal uses.

As mentioned earlier, many public agencies, including all the federal agencies, will contract for seed to be collected from the land they manage and use it onsite, or have the collected seed increased in a production field prior to use. The NPS almost exclusively uses this approach for park restoration projects. However, some agencies need a very large quantity of seed for restoration projects, particularly after wildfire, so the largest land-management agencies turn to commercial suppliers. For perspective, in 2020, BLM field offices purchased about 1.5 million pounds of seed (a little less than in a typical year) of which two-thirds was used for post-fire seeding on over 422,000 acres of burnt land. BLM has two warehouses for storing purchased seed on a short-term basis.

Field offices are encouraged, but not required, to request native seed genetically adapted to where it will be planted. In 2020, about one-eighth of the grasses, one-fourth of the forbs, and two-thirds of the shrubs purchased were native seed of certified (source-identified) origins. Another portion of the purchased seed (particularly of grasses and forbs) included native seed types more available in the market and at lower relative cost, but that originate from plant communities in geographic locations mismatched to the environments where the seed is likely to be planted. Similarly, non-native plant seeds, that are less expensive and more available, were a sizable portion of the seed purchased.

TABLE S-1 Uses of Native Seeds by Federal and State Agencies

Creation or restoration of wildlife habitat (other than pollinator habitat)

Pollinator habitat projects

Stream erosion mitigation or restoration

Restorative activity on land in a natural area or wilderness

Soil protection

Invasive species suppression

Roadside seeding and maintenance

Landscaping

Green infrastructure (bioswales)

Natural disaster recovery

Rangeland grazing

Energy development remediation

Green strips (vegetative fuel breaks)

Recently, the USFS announced plans to address the fire-prone condition and reforestation backlog across the National Forests. The USFS will use funds of up to $123 million annually[4] to reforest over 4 million acres in the next 10 years. The Service's six nurseries will produce tree seedlings and other plants, but private growers will be needed to expand the capacity for production. The USFWS, NPS, DOD, and most states have at least one nursery to grow native plants or trees for use in restoration. Several of the tribal nations have established nurseries, and more want to, but efforts to expand a network of tribal nurseries have been limited by lack of resources.

Filling the Native Seed Pipeline

The ability of the commercial sector to provide the seed needed for projects in different landscapes is important, because existing native plant communities do not produce the same amount of seed each year and can be degraded by over-collecting. The committee's survey of state agencies and presentations by federal agency personnel suggest that buyers often settle for substitutions because their preferred seed is simply not available at any price during the timeframe in which the seed is needed. Timeframe is an issue for BLM field offices that rely on annually appropriated funds for post-fire needs that must be spent before the end of the federal fiscal year, forcing them to buy whatever is available.

Seed suppliers in the committee's survey reported that the greatest challenges they face in supplying native seed are unpredictable demand, "difficult to grow" species, and a lack of stock (starter) seed from appropriate seed transfer zones (seed zones, for short), which are geographically mapped zones inside of which plants can be relocated and are presumed to be genetically adapted. They suggested that better communication from buyers about their seed needs, more realistic timelines for delivery that account for the time needed to acquire and propagate seed, and greater technical assistance in growing native plants would help them meet user needs.

In 2018, BLM's Plant Conservation and Restoration Program began to address the issues of communication, timeframe, and availability of stock seeds using a native seed Indefinite Delivery Indefinite Quantity (IDIQ) production contract with suppliers to have them grow and increase native seed types from seed transfer zones not readily available on the commercial market. Although seed production IDIQ contracts are not a new financial instrument, the focus on seed production by seed transfer zone is a new approach for BLM. The seed given to suppliers for increase came from the Seeds of Success (SOS) program initiated by BLM in 2001. To secure the genetic diversity of native plants, BLM partnered with six nonfederal seed banks (SOS Partners)[5] to store native seed collected from numerous seed zones across different ecoregions (see Figure S-1), mostly in the western United States. Although the quantities being increased are small and the funding for the new program is modest, the IDIQ approach is a significantly new approach to meeting seed needs.

State and Regional Partnerships

Because locally adapted native seed for a wide range of species has been difficult to find in the commercial marketplace, cooperative partnerships have arisen inside and outside of the federal government to develop native seed supplies for restoration and other applications, including to provide stock seed to commercial growers. For example, in the 1980s, the Iowa Tallgrass Prairie Center partnered with the Iowa Department of Transportation to collect seed from the state's remaining stands of tallgrass prairie found along rural roadsides. The program began increasing the seed in fields and has since released stock material for 100 species of Iowa natives to commercial suppliers. In addition to prairie restoration projects, the seeds are now used on roadsides across the state, and are available to farmers and other landowners who participate in the USDA-supported Conservation Reserve Program, which puts marginal farmland into conservation use for a decade or more. The approach offers one model for supplying and developing a native seed industry on a state level.

Several regional native plant materials development and restoration programs have been established by BLM and the USFS that are broadly cooperative, involving government at all levels, the tribes, colleges, nongovernmental

---

[4] The funds will be available through the Repairing Existing Public Land by Adding Necessary Trees Act, which was included in the 2021 Infrastructure Investment and Jobs Act, commonly referred to as the Bipartisan Infrastructure Law.

[5] Text was added after the prepublication release to clarify that the seed banks store the seed for the SOS program; they are not the seed suppliers that increase the seed.

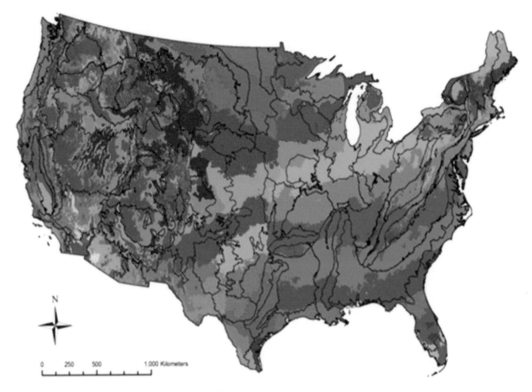

FIGURE S-1 Seed zones overlaid with ecoregions.[6]
NOTE: Generalized provisional seed zones (colored areas) are overlain by Omernik Level III Ecoregions.
SOURCE: Based on Bower, A.D., J.B.S. Clair, and V. Erickson. 2014. Generalized provisional seed zones for native plants. Ecological Applications 24(5):913–919 and US Environmental Protection Agency. 2013. Level III ecoregions of the continental United States: Corvallis, Oregon, National Health and Environmental Effects Research Laboratory.

organizations, and native seed producers. These include programs situated in the Pacific Northwest, the Colorado Plateau, the Great Basin, and the Mojave Desert that collect and bank seeds, release seeds for commercialization, conduct landscape genetic studies, and carry out research on restoration techniques and strategies, among other things. Some of those efforts involve more than 30 partner organizations and demonstrate what public-private partnerships can achieve working together toward focused goals.

Such partnerships, developed on a regional basis, may have been envisioned in the plan given to Congress in 2002 in response to its challenge to build a seed supply, but the native seed supply requires transformation on a broader scale. In 2015, the National Seed Strategy, a plan "to foster interagency collaboration to guide the development, availability, and use of seed needed for timely and effective restoration," was released by the Plant Conservation Alliance, which now includes more than 400 nonfederal organizations and 17 federal agencies.[7] That Strategy is not a funded program, but under its umbrella many diverse projects have been pursued by different agencies and their partners. It is a sound blueprint for native seed development and continues to be a call for action.

Developing a sustainable, national native seed supply is not likely to be achieved for many years, however, because the ongoing efforts are focused beneath the ambitions of that goal. Moreover, current efforts have been outpaced by needs that have escalated very quickly over the last decade as the result of converging threats to

---

[6] In 1987, James Omernik published a map of ecoregions across the United States defined by major differences in geography and climate [Omernick. 1987. Ecoregions of the Conterminous United States, Annals of the Association of American Geographers, 77(1): 118–125 doi:10.1111/j.1467-8306.1987.tb00149.xe]. Subsequent refinements by the Environmental Protection Agency led to the development of more discrete ecoregions. In Figure S-1, the boundaries of ecoregions are shown by the black lines.

[7] The text was corrected after the prepublication release to update Plant Conservation Alliance numbers.

natural ecosystems. Although the federal agencies collaborate productively with one another and other partners on a project-by-project basis, they could have a greater impact if they turned their focus to the bigger picture and committed to a unified agenda to develop a native seed supply that meets the nation's diverse needs. Now more than ever is an opportunity to pursue that goal, with the recent passage of the Infrastructure Investment and Jobs Act (Bipartisan Infrastructure Law) (2021),[8] the Great American Outdoors Act (2020),[9] and fresh new thinking about the nation's natural resources, such as through the America the Beautiful initiative (2021).[10]

SUMMARY CONCLUSIONS

The following conclusions summarize the committee's investigation of the experiences of federal land-management agencies, state governments, tribal nations, collaborative partnerships, and suppliers.

Conclusion 1-0: *There is urgency to building a native seed supply for the restoration of native plant communities. Developing reliable seed supplies for ecological restoration is an achievable goal by the federal agencies, but one that demands accelerated, inter-institutional commitment to a comprehensive vision, at a much more intensive level than is currently under way.*

Conclusion 2-0: *In some regions (at, above, or below the state level), native seed needs are being addressed by networks or partnerships that include federal agencies, states, tribes, local governments, and the private and nonprofit sectors. Further development of these regional networks, plus greater coordination among them, is a promising way to stabilize demand, expand supply, and increase the sharing of information and technology that is critical to meeting native seed needs.*

Conclusion 3-0: *Tribal uses for native plants parallel those of federal and state agencies and their needs include seed collection, increase, seed testing, implementation of seed zones, storage, and contracting for ecological restoration. The historic complexity of land-management issues and interactions with federal and state governments, former Bureau of Indian Affairs policies associated with cultural assimilation, and a lack of resources have constrained all aspects of tribal land management including activities to build capacity to meet native plant needs.*

Conclusion 4-0: *Suppliers view unpredictable demand as their leading challenge. Suppliers indicate that seed contracts with adequate lead time, clear delivery timelines, price guarantees, and a guarantee to purchase a predetermined quantity of seed of the specified ecotypes are most important to them.*

Conclusion 5-0: *Suppliers indicate that difficult-to-grow species and lack of stock seed from appropriate seed zones or locations are their top two technical challenges. They identified communicating demand (issues related to planning, communication, funding, and the economics of the seed markets) as key to helping them achieve success in providing seeds to buyers.*

Conclusion 6-0: *Many technical and scientific information gaps affect the ability of the native seed supply to function efficiently and effectively. Addressing them would inform decision making, reduce uncertainty, and improve restoration outcomes.*

Conclusion 7.0: *To enable more producers to enter the supply chain for native, ecoregional seed of diverse species, there is a need to expand cooperative seed cleaning facilities and humidity-controlled seed warehouses in areas where commercial facilities with specialized equipment do not exist.*

---

[8] H.R. 3684, Infrastructure Investment and Jobs Act, see https://www.congress.gov/bill/117th-congress/house-bill/3684 (accessed February 9, 2023).

[9] H.R. 1957, Great American Outdoors Act, see https://www.congress.gov/bill/116th-congress/house-bill/1957 (accessed February 9, 2023).

[10] America the Beautiful, see https://www.doi.gov/priorities/america-the-beautiful (accessed February 9, 2023).

Conclusion 8-0: BLM is the nation's largest user of native seed for restoration and has the largest capacity for seed storage, but many constraints currently limit its capacity to act as a reliable purchaser of native seeds and thereby to support a more robust native seed industry. The capacity and staffing of its seed warehouses are also inadequate to meet existing and projected needs.

Conclusion 9-0: BLM manages extensive areas of natural and seminatural land that constitute an important in situ repository of seed stock for restoration, but conserving this critical resource is not yet recognized as a land-management objective. The agency's native seed needs depend on the native plant communities that are the source of seeds, adding to the urgency for recognizing and protecting native plant communities and biodiversity on BLM lands.

Conclusion 10-0: Developing reliable seed supplies for restoration is an achievable goal for BLM and other public-sector users of native seed, but one that demands substantial institutional commitment to make restoration a high-priority objective, the cultivation and empowerment of botanical and ecological expertise in decision making, and the creation of sustainable funding streams for restoration that enable long-term planning, successful implementation, and learning from experience.

RECOMMENDATIONS

Recommendation 1.0: The leadership of the Departments of the Interior (DOI), Agriculture (USDA), and Defense (DOD) should move quickly to establish an operational structure that facilitates sustained interagency coordination of a comprehensive approach to native plant materials development and restoration.

An interagency approach focused on fulfilling the long-standing congressional mandate to develop a native seed supply for public lands could unify the agencies' independent efforts to meet seed needs for restoration and rehabilitation. This focused effort would maximize returns on investment in native plant materials development and restoration and would augment existing activities within agencies. One possible model for agency coordination on a national basis, organized regionally, is the National Interagency Fire Center.

Among the activities that could be the focus of interagency efforts are the following:

- Serve as the central coordinating platform for developing a national native seed supply.
- Assist the launch, support, and oversight of regional native seed supply development activities.
- Coordinate the prioritization of species and ecotypes to meet seed needs for different regions.
- Co-develop national policy for native seed collection, seed sharing, and seed use.
- Co-develop and share best management practices for seed choice in restoration.
- Coordinate, prioritize, and support basic and applied research such as described in Chapter 8 and other region-specific research needs identified in the National Seed Strategy.
- Review and strengthen policy guidance for the use of native seeds on public lands.
- Co-develop adaptive management approaches that use experimentation during restoration, gather data on outcomes, and use these data to guide future restoration.
- Provide a national, central data-collection platform and analytical capability.
- Serve as a focal point for training on seed collection protocols, storage practices, seed cleaning and testing, and other technologies.
- Produce information and technology for stakeholders in the native seed supply and native plant restoration.

Recommendation 2.0: Federal land-management agencies should participate in building regional programs and partnerships to promote native plant materials development and native plant restoration, helping to establish such regional programs in areas where they do not yet exist. Ideally, the existing regional programs and partnerships would grow into a complete nationwide network, assisted by the federal interagency coordinating structure envisioned in Recommendation 1.0. The size, geographic coverage, and membership of the regional programs would vary based on regional needs and would include many existing entities such as the USDA Plant Materials Centers and Agricultural Research Service (ARS) seed banks, other seed banks, botanic gardens, public nurseries, universities, and other organizations.

2.1 Develop seed priorities. Each region has its workhorse native plant species, ones that are easily grown and highly abundant in natural communities, as well as its other priority species such as those that are important for pollinators and wildlife. Developing region-specific lists of priority species will enable suppliers to focus on developing stocks of the species, and ecotypes of these species, that are likeliest to be in high demand.

2.2 Conduct scenario planning and monitoring. To enable suppliers to anticipate and meet future needs, scenario planning at the regional level should be used to estimate the kinds and quantities of seeds likely to be in demand over multiyear horizons. Such planning scenarios would be based on estimates of current degraded land and future events likely to cause further degradation. These would be periodically updated through ongoing monitoring.

2.3 Collect and curate stock seed. Regional programs should conduct, or oversee the conducting of, the collection and curation of wildland-sourced seed to make it available for future production. Seed must be collected using protocols to record source locations, maximize genetic diversity, and protect wild populations. As banked stocked seed accessions deplete, additional wild collections will be necessary. As source populations are lost, or better ones located, ongoing monitoring and reevaluation will be needed.

2.4 Share information. Regional programs should develop informational tools derived from monitoring regional seed use practices and the success or failure of outcomes. This will inform smarter, more predictable selections of species for procurement in the regional marketplace and more effective restoration strategies.

Recommendation 3.0: The Bureau of Indian Affairs should work with the Inter-Tribal Nursery Council to promote and expand tribal nurseries. The Bureau of Indian Affairs should prioritize and support tribal native plant uses and capacities, consistent with the Bureau's legal and fiduciary obligation to protect tribal treaty rights, lands, assets, and resources held under the Federal-Tribal Trust. Recognizing tribal sovereignty and self-determination, expanding the cultural, economic, and restoration uses of native plants by tribes will require the promotion and expansion of tribal nurseries and greater support for the Inter-Tribal Nursery Council. Additionally, tribal leaders and land managers should be fully engaged in planning, conducting, and applying results from scientific projects related to seed production and conservation, native plant restoration, and ecosystem management on tribal land.

Recommendation 4.0: The public agencies that purchase native seed should assist suppliers by taking steps to reduce uncertainty, share risk, increase the predictability of purchases, and help suppliers obtain stock material.

4.1 Conduct proactive restoration on a large scale. Millions of acres of US public land are ecologically impaired. With new federal resources for restoration, federal and state agencies should plan restoration projects on a 5-year basis, ensure that stock seed has been made available to suppliers, and set annual purchase targets for the collection and acquisition of needed ecotypes of native plant species. These actions will result in considerable expansion and stabilization of the market for native seeds, benefiting suppliers and users alike.

4.2 Establish clear agency policies on native seed uses. Land-management agencies should establish clear policies on seed use on lands under their stewardship that support the use of locally adapted native plant materials in management activities, along with clearly delimiting the circumstances for allowing exceptions. This will send a strong signal of species and provenance needs to suppliers.

4.3 Support responsible seed collection and long-term seed banking. Intensive and carefully managed seed collection is needed to supply the native plant material enterprise and conserve native plant diversity for the long term.

In cases where native seed is collected from public lands by private suppliers for direct sale and use in restoration, land-management agencies should employ adequate personnel to issue permits and ensure responsible collection.

In other cases, where seed is collected for increase and native plant materials development, the federal agencies should facilitate this activity by extending the Seeds of Success program to include all regions of the United States, and better supporting its activities.

In still other cases, native seed is collected and banked for long-term conservation. To continue building a species-diverse and genetically diverse long-term native seed bank, the federal agencies, led by BLM and the ARS National Plant Germplasm System, should accelerate their collaboration under the Seeds of Success program.

4.4 Contract for seed purchases before production. Public-sector buyers of native seed should amend their policies to enable contracting for purchases before the seed production cycle begins. Forward contracting, in which the buyer agrees to purchase specific seed at a future date, reduces grower risk by ensuring them a market.

4.5 Use marketing contract features that reduce demand uncertainty. Public-sector buyers of native seed should adopt marketing contract features that specify the delivery timeline, guaranteed prices, and guaranteed purchase of predetermined quantities that meet the buyer's specifications.

4.6 Experiment with native seed contract designs. Public-sector buyers of native seed should experiment with cost- and risk-sharing contract designs, such as BLM's IDIQ contracts, which are partially supported by a working capital fund.[11] Other approaches also merit investigation, including Blanket Purchase Agreements. In special circumstances, such as where high-priority species or ecotypes are unavailable, federal agencies should consider issuing contracts to suppliers for research and development.

4.7 Consider providing premiums for local ecotype use in USDA conservation programs. USDA's Farm Service Agency and state departments of agriculture should explore whether higher co-payments for landowners' use of local ecotypes of native species in conservation program plantings would enhance environmental benefits while supporting a regional native seed industry. A requirement for the use of certified seed would assure growers that their efforts and expenses are not being undercut by seeds of unverified origins.

Recommendation 5.0: Federal land-management agencies should work with their regional partners to launch an outreach program to provide seed suppliers with critical tools and information.

5.1 Strengthen the role of the National Resources Conservation Services (NRCS) in supporting native seed suppliers. The NRCS Plant Materials Centers can build on their historical role by developing materials and techniques and advising commercial growers on native seed production, emphasizing non-manipulated germplasm and production methods that minimize genetic change.

---

[11] The text of the recommendation was modified after release of the prepublication to clarify that the BLM IDIQ is not intrinsically tied to the working capital fund and can be supported from other funding sources.

5.2 Support information sharing on restoration outcomes. Public land-management agencies should share new research and technical knowledge on restoration in publicly available technical progress reports.

5.3 Facilitate communication with growers. To support existing native seed growers and to encourage new ones, federal agencies should provide tools such as an online marketplace, and resources such as workshops and information on propagation, seed cleaning, and other techniques.

Recommendation 6.0: The federal government should commit to an expanded research and development agenda aimed at expanding and improving the use of native seeds in ecological restoration.

6.1 Support basic research. Basic research to support restoration in an era of rapid change should be a priority for the National Science Foundation, USDA National Institute of Food and Agriculture, and the US Geological Survey (USGS). Some critical topics include restoration under rapid environmental change, species selection to promote ecological function, traditional ecological knowledge, and the economics of the seed market.

6.2 Build technical knowledge. Development and dissemination of new technical knowledge for restoration should be important priorities for the research arms of federal land-management agencies, such as the USDA-ARS, USFS, DOD, and USGS. Priority topics include improving techniques for production, maintenance of genetic integrity and quality, testing, storage, and deployment of native seeds.

6.3 Adaptive management. Public land agencies and their partners should commit to using a rigorous adaptive management approach that documents all features of the restoration plan, uses restoration treatments as experiments to address critical areas of uncertainty, gathers data on outcomes, and uses these data to guide future restoration actions.

6.4 Seed zones. USDA-ARS, USFS, and USGS, in conjunction with regional programs, should seek to develop a more uniform national system of seed zones, which will be an important step in sending clearer signals to both seed suppliers and seed users.

Recommendation 7.0: Federal agencies and other public and private partners, including seed suppliers, should collaborate on expanding seed storage and seed-cleaning infrastructure that can be cooperatively cost-shared regionally. Additional storage can improve the availability of seed ready for restoration when urgent but hard-to-predict needs arise. Greater refrigerated and freezer storage availability would protect the viability of purchased seed until its use. Seed cleaning is also a significant technical challenge for new and small-scale suppliers. Regional native seed-cleaning facilities would encourage more growers to enter the supply chain.

Recommendation 8.0: BLM's Seed Warehouse System needs to be expanded, particularly its capacity for cold storage, and supported by staff with up-to-date knowledge of seed science to manage the seed inventory. The BLM seed warehouses belong in a national program within BLM or DOI to ensure they are funded, well managed, expanded to meet national needs, and shared among agencies. With more warehouses, seed can be kept near the location where it will be used, reducing transportation costs, time, and seed viability loss.

Recommendation 9.0: BLM should identify and conserve locations in which native plant communities provide significant reservoirs of native seeds for restoration. Public land-management agencies should actively recognize and protect the natural plant communities that provide the ultimate sources of native seeds for ecological restoration, using protective designations such as Area of Critical Environmental Concern or Research Natural Areas.

Recommendation 10.0: The Plant Conservation and Restoration Program (PCRP) should be empowered with the capacity to plan and oversee restoration and to build stocks of seed. BLM should give greater authority and resources to the PCRP to oversee seed purchasing and warehousing decisions, monitor restoration outcomes, engage in long-term restoration planning, and ensure that staff with plant expertise are available to guide restoration planners and seed purchasers across BLM. This expanded role for PCRP will hasten the pace of change toward the use of natives that is already under way. The PCRP should expand the use of IDIQ or other innovative, risk-sharing contracts to build a diverse supply of native seed in BLM warehouses.

FINAL THOUGHTS

These recommendations represent an ambitious agenda for action, commensurate to the challenges facing our natural landscapes, and to the responsibility for public-sector leadership of a coordinated public-private effort to build a national native seed supply. The committee is optimistic that the many public and private parties engaged in ecological restoration across the nation will be willing partners in ensuring this agenda's success.

1

Introduction

In 2020, the National Academies of Sciences, Engineering, and Medicine released *An Assessment of the Need for Native Seeds and the Capacity for Their Supply: Interim Report* (NASEM, 2020), completing the first of a two-phase study of the national need for and supply of native seeds for ecological restoration and other purposes. The National Academies appointed an ad hoc committee of experts to conduct the study at the request of the Bureau of Land Management (BLM), which needs large quantities of native seed to meet its public land-management responsibilities, particularly in the aftermath of wildland fires.

The interim report provided an overview of the participants in the native seed supply chain in the United States, made preliminary observations about the dynamics of the seed supply, and proposed an information-gathering strategy for the second phase of the study to obtain deeper insights into the nation's capacity for providing the seeds needed by users with various objectives.

This final report presents findings of the information-gathering process carried out in 2021 and 2022. During this period, the committee sought information from, and about, various actors in the native seed supply chain including federal, state, tribal, and municipal agencies, private landowners in conservation programs, native seed collectors and suppliers, seed testing associations, seed banks, land trusts, environmental groups, and other nongovernmental organizations.

Using surveys and interviews, presentations to the committee, expert consultations, and the considerable experience of its own members, the committee explored the diverse set of needs and activities in the native seed supply chain to identify opportunities for progress. Based on its findings and conclusions, the committee offers a set of recommendations for increasing the supply of native seeds to meet the increasing and evolving demand for native seeds.

STATEMENT OF TASK

To guide the assessment, the committee was given a statement of task (Box 1-1) which focused the committee on a list of straightforward questions: What entities use native seeds, how do they use them and at what scale, what shapes their decisions, how do seed buyers and suppliers communicate, and what is the nature of procurement? How does seed availability for restoration relate to other agricultural, land-management, and conservation activities? What activities make up the seed supply chain and how well are they working to meet seed needs?

Simple answers to these questions were not as straightforwardly obtained, however, and there were limits to the committee's ability to obtain a complete picture of the native seed supply. In addition, the questions in the

---

BOX 1-1
Statement of Task

An ad hoc study committee appointed by the National Academies of Sciences, Engineering, and Medicine will assess federal, state, tribal, and private-sector needs and capacity for supplying native plant seeds for ecological restoration and other purposes. The assessment will focus on the western continental United States and incorporate information from assessments of other US regions, as available, toward the goal of a nationwide perspective. The assessment will be carried out in two phases. In phase one, the committee will conduct fact-finding, develop a framework for information gathering for the assessment, and prepare an interim report describing the framework and implementation strategy. In the second phase, the committee will oversee the data- and information-gathering process, analyze the information obtained, and prepare a final report summarizing the committee's findings and conclusions. The final report also will provide recommendations for improving the reliability, predictability, and performance of the native seed supply.

The assessment may include information on

- how native seeds are being used by public (federal, state, tribal) and private (land trusts, companies, and nongovernmental organizations) in ecological restoration and other activities;
- the frequency and scale of the demand and the characteristics of the seeds pursued by users, as well as the diversity of applications for which they are sought;
- how users find seeds that are appropriate for their intended purpose and how users communicate their needs for seeds to potential suppliers;
- how suppliers make known their capacity to potential users;
- the different kinds of entities and roles that compose the seed supply chain (from professionals and organization involved in the identification of site-specific needs, to the collection, propagation, cleaning, storage, and supply of seed) and their respective capacities;
- the relationship of seed availability to other agricultural, land-management, and conservation activities generally;
- procurement processes for native seeds and the cost, availability of funds, infrastructure, market, and other factors that influence decision making on the part of users and suppliers of native seeds;
- opportunities to increase the size and capacity of the native seed supply chain (and number of suppliers); and
- other relevant issues identified by native seed users and suppliers and other stakeholders.

---

statement of task belie considerable nuance and complexity, beginning with the term "native seed,"[1] which, in addition to its natural history, can be defined by many distinguishing characteristics that are important to seed buyers who plan to use them in restoration projects. Apart from the fact that native seed in the United States comprise a diversity of species and of forms (trees, shrubs, grasses, forbs), factors such as where exactly the seed or its progeny was sourced, the populations from which it was collected, and the way the seed was developed into the final product offered for sale vary significantly.

---

[1] The statement of task (see Box 1-1) specifically refers to "native plant seed." The report uses native seed as a shorthand term for seed and all other "native plant materials" such as containerized plants (see Glossary). These terms are used separately in Chapters 4 and 7, in the results of surveys in which questions were asked specifically about the use or production of either native seed or native plant materials (such as nursery stock). Native seed is obtained from terrestrial and aquatic plant species that have evolved and occur naturally in a particular region, ecosystem, or habitat. Species native to North America are generally recognized as those occurring on the continent prior to European settlement (PCA, 2015).

Buyers of native seed, whether in the public or private sector, are not monolithic in their preferences for seed characteristics. Describing a "need" for native seed therefore requires a frame of reference relative to the view of the user with respect to the activities in which seed will be used. They might describe their "needs" for seed in terms of soil stabilization, creation of pollinator habitat, or stormwater mitigation. Native plants are regarded as inherently valuable for providing essential ecosystem goods and services that support users' objectives, but users may not consider their objective as one of ecological restoration. They may be willing to use native plants manipulated by breeding or selection which would not be acceptable to others or for other applications (or even qualify as "native" by some users).

An important frame of reference for the assessment are the needs of federal agencies such as BLM and the US Forest Service, and those of some state agencies, charged with the mission of managing public lands for multiple uses, like grazing, recreation, wildlife habitat, and hunting, and for responding to the effects of major disturbances, such as wildfire. These agencies use thousands of pounds of native seed every year for these activities. Because agencies have finite resources to address their land-management responsibilities, the "needs" for native seed are partially defined by priorities handed to them—first, to address post-fire emergency stabilization; second, to support ongoing priorities (sage-grouse habitat) and other multiple uses of the land; and finally, to address restoration and conservation needs. Decision making in large agencies is distributed, and the basis of seed choices and outcomes of seed use are not routinely documented, thus making it challenging to pin down users' perception of needs. The point here is that seed needs are shaped by a broader context.

Preliminary Observations as the Context for Phase Two

The committee's interim report included eight preliminary observations about the native seed supply (Box 1-2), which were used as an informal set of hypotheses that shaped the questions developed for the information-gathering phase. In this respect, they provide a useful context for the findings from all the chapters but particularly semi-structured interviews with federal agency personnel (Chapter 3), and the results of the survey of state agency personnel (Chapter 4), and native seed suppliers (Chapter 7).

BOX 1-2
Preliminary Observations about the
Native Seed Supply from the Interim Report

Observation 1. Users of native seed have varied objectives and needs.
Observation 2. The seed market in the western United States is strongly affected by decision making by the large land-management agencies, such as the Bureau of Land Management (BLM) and US Forest Service (USFS).
Observation 3. Timeframe, quantity, and quality strongly limit the overlap between what seed is available and what seed is desired.
Observation 4. Seed choices do not always support restoration success, and outcomes do not always inform choices.
Observation 5. The budgets and seed specifications of users vary greatly, as do the unit costs of suppliers.
Observation 6. Seed procurement may be hampered by market volatility, risk, and contract structure.
Observation 7. The seed market may be strongly affected by a limited capacity for seed banking and for adequate and appropriate storage conditions.
Observation 8. Issues affecting urban, mid-western, and eastern settings are somewhat distinct from those affecting public lands in the West.

BROADER CONTEXT FOR THE ASSESSMENT

It is important that readers of this report be aware of a broader context for the native seeds assessment that includes the historical efforts to build a native seed supply, and the high stakes, which some might argue, are existential in nature, to achieving that goal, as well as recent opportunities that offer an optimism for success. Described here, they played a role in shaping the committee's deliberation for actions that might be taken to supply native seed needs.

The Congressional Mandate for a Native Seed Supply

In June 2001, with increasing acres of public land affected by wildland fire, Congress urged the Department of the Interior's (DOI) Burned Area Emergency Rehabilitation program to go beyond emergency stabilization and toward the rapid use of native plant species to prevent invasive species encroachment in newly burned areas. In the committee report accompanying the House 2002 Appropriations Bill for DOI and Related Agencies, the Secretaries of the Interior and Agriculture were directed to report jointly to Congress by the end of 2001, "with specific plans and recommendations to supply native plant materials for emergency stabilization and longer-term rehabilitation."[2]

The response to Congress was prepared by BLM and three other land-management agencies, the US Forest Service (USFS), National Park Service, and the US Fish and Wildlife Service, with input from the US Department of Agriculture's (USDA's) Natural Resources Conservation Service (NRCS) and Agricultural Research Service, along with the DOI's US Geological Survey, and the Office of Surface Mining. The plan conceived by the interagency group put forward three elements key to the success of a long-term program for native plant materials development:

1. Support for Federal, State, and Tribal Production, Development, Storage, and Research Facilities
2. Public-Private Partnerships
3. Education and Outreach

The report to Congress (DOI, USDA 2002) cautioned that the infrastructure to build a native seed supply was in its infancy but pointed to the achievements of the past as evidence that the nation could address the challenge. The report noted the history of successful plant material development by USFS nurseries in the 1920s with conifer tree seedlings and by NRCS Plant Materials Centers in the 1930s Dust Bowl era with plant species for conservation.

The 17-page plan put forward five action items:

1. Undertake a comprehensive assessment of needs for native plant materials.
2. Make a long-term commitment to native plant materials production, research and development, education, and technology transfer.
3. Expand efforts to increase the availability of native plant materials.
4. Invest in partnerships with state and local agencies and the private sector.
5. Ensure adequate monitoring of restoration and rehabilitation efforts.

Twenty years later the plan submitted to Congress in 2002 seems rational, but the continued shortages of native seed suggest that its aspirations are not yet realized. The final paragraph of the 2002 report to Congress alluded to the potential for inertia, mentioning the need to recognize the "different missions" of agencies and the reality that the infrastructure needs of one agency "are not necessarily shared" by others. The National Seed Strategy developed in 2015 by the Plant Conservation Alliance, a coalition of public and private partners dedicated to native plant conservation, included the establishment of a Memorandum of Understanding between 12 federal agencies to work that signified a willingness to coordinate separate ongoing activities of the agencies and to pursue

---

[2] House Committee on Appropriations Report, to accompany H.R. 2217, Department of the Interior and Related Agencies Appropriations Bill, 2002, 107th Congress.

opportunities for cooperation. There are many productive activities taking place under the umbrella of the Strategy, but not necessarily at the scale needed to meet the needs. The next section of the report provides a lens through which the high stakes of this national effort should be viewed.

The Dual Crises of Biodiversity Loss and Climate Change

As part of an ongoing human-mediated biodiversity crisis, organisms face many threats, not only fire, but water and air pollution, climate change, the negative impact of introduced species, and disease. The destruction of natural habitats in the United States as well as worldwide is a major threat to the survival of species (Pimm et al., 2014; International Union for the Conservation of Nature[3]). Biodiversity loss is a major threat to ecosystem function and human well-being. One widely accepted estimate is that extinction rates (largely human mediated) are now 1000 times higher than what would be considered a typical background rate (De Vos et al., 2015; Pimm et al., 2014). Similarly, the "broad footprint" of climate change ranges from alterations in fine-scale genetic variation to entire ecosystems, with impacts on every ecosystem on the planet (Scheffers et al., 2016).

The assessment of native seed needs and capacity has taken place over sequential years in which climate change has become a major influence on the environment. The release of the interim report in October 2020 coincided with one of the most expansive western wildland fire seasons on record, with large fires consuming more than 10 million acres (NICC, 2021). That summer was also the most active Atlantic hurricane season of all time, with 7 major hurricanes out of 14 total in 2020 (NOAA, 2021). In 2021, the western United States and Canada experienced a record-breaking heat wave unprecedented in the region's history, with temperatures above 113 degrees F for consecutive days, and during which British Columbia recorded a record high of 115.8 degrees F. This year, 2022, the long-term drought in the western United States continued through its 23rd year, a time span of dry conditions not believed to have occurred since the year 800 AD (Williams et al., 2022), while deadly flash floods in the spring and late summer swept across Yellowstone, Las Vegas, St. Louis, Kentucky, and Death Valley.

These extraordinarily severe events reflect a widespread shift in the norm of weather patterns. Extreme events can and do damage landscapes; wildfire is after all one reason public land managers seek native seeds. Intact native plant populations have evolved to be resilient, however, and are thought to be able buffer the impacts of climate change on ecosystems and support recovery of animal species. Land-management priorities have not always considered the fact that native plant communities are a key natural resource underpinning different uses of the land. Thus, conserving and restoring the integrity of native plant communities has emerged as an important, if neglected, ecological "need," particularly on the millions of acres of public lands experiencing environmental degradation.

From a land-management standpoint, conserving and supporting the integrity of existing plant communities across environmentally diverse landscapes is relevant to climate adaptation and mitigation, and gives additional urgency to the conservation of plant communities as an important genetic resource and a source of seed to assist plant community restoration elsewhere in the future (Havens et al., 2015).

As a final point of perspective, the Seeds of Success[4] (SOS) program initiated by BLM in 2000 to collect seed from wild stands of native plants nationwide is estimated to have collected and placed in storage approximately one-third of the nation's native plant biodiversity (5,600 species of 18,000). A recent analysis of the location of fire incidence relative to SOS collection in the western United States documents where collection sites have been burned. It predicts that 14 percent of the sites of SOS collections will have burned by 2050 if fire frequency follows the trajectory it has taken since 2011 (facilitated in part by the spread of invasive grasses that support an accelerated fire regime). While recovery after fire is possible, the stored collections may be the only source of seed representing the genetic variation if wild stands are lost (Barga et al., 2020).

---

[3] See www.iucn.org.

[4] See https://www.blm.gov/programs/natural-resources/native-plant-communities/native-plant-and-seed-material-development/collection.

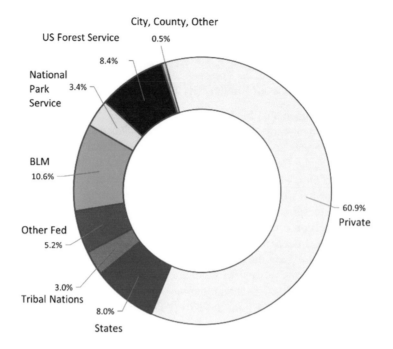

FIGURE 1-1 Land ownership in the United States.
SOURCE: Adapted with permission from Headwaters Economics, based on US Geological Survey, Gap Analysis Program 2018. Protected Areas Database of the United States, Version 2.0.

Land Ownership and Native Seed Needs

Land ownership is also a relevant lens for the assessment. The committee focused largely on understanding of public-sector (state and federal) seed needs. As shown in Figure 1-1, almost 40 percent of the 2.3 billion acres of land in the United States is under public ownership or management. Federal agencies control 27.1 percent of US land, or about 640 million acres. The other major public landowners are state governments (8.0%) and tribal governments (3.0%).

Ninety-two percent of all the US land under federal control is in the 11 western states of the contiguous United States and Alaska (Figure 1-2), and with Hawaii, amounts to almost 51 percent of all those states combined (Vincent et al., 2020). Important native plant resources are distributed across the intertwined land holdings under federal, state, tribal, and private ownership.

Preliminary Observation 2 of the committee's interim report postulated that the native seed market in the western United States is strongly affected by decision making by the large land-management agencies, such as BLM and USFS. The survey of state agency personnel partially supported this hypothesis, as seed shortages were associated with bad fire years, but another federal program, the USDA Conservation Reserve Program (CRP) was also shown to play a strong role in how suppliers anticipated the demand for native seed (see Chapter 7). That may be because 63 percent of privately owned land in the nation is occupied by farms and ranches, and CRP funding produces a relatively consistent demand for seed mixes with stock material for production available to seed suppliers.

In eastern states, there are fewer federal lands (4% of total land) and a greater percentage of state-managed land. The amount of state land differs by state (for example, New Jersey, 21%; Florida, 16%; Pennsylvania, 14%; Michigan, 13%; and Minnesota, 11%). The assessment could not discern if there are more suppliers in the West than in the East because of a federal presence to support a larger industry, but observations about the federal use of native seeds and their impact on the supply chain may differ in the eastern states.

FIGURE 1-2 Federal lands map.
SOURCE: Geographic Information Systems (GIS) Geography 2022. See https://gisgeography.com/federal-lands-united-states-map/ (accessed on February 9, 2023).

New Opportunities for Ecological Restoration

A final contemporary context for the native seed assessment is the significant development during the past 2 years of national directives that recognize the value of natural assets in relation to jobs and the national economy. At no time in recent legislative history has the US Congress been more supportive of calls to protect natural areas and their biodiversity. Bills passed in 2021 and 2022 include the Infrastructure Investment and Jobs Act (Bipartisan Infrastructure Law[5]), which includes $200 million for the National Seed Strategy ($70 million for the DOI, and $130 million to USDA) in the next 5 years, as well as the Repairing Existing Public Land by Adding Necessary Trees (REPLANT) Act, providing up to $140 million annually for reforestation on National Forests. The Great American Outdoors Act[6] will provide up to $1.9 billion for the next 4 years to maintain infrastructure at national parks, forests, wildlife refuges, and recreation areas, and $900 million in annual funds to the Land and Water Conservation Fund to allow federal and state acquisitions of recreational land. The current executive administration has put forward ambitious plans for nature-based solutions to climate change (America the Beautiful[7]) and to take stock of the contribution of natural capital to the national economy (Earth Day Executive Order for a natural capital accounting[8]). With the resources allocated by these initiatives, federal, state, and tribal agencies have an unprecedented opportunity to conserve and restore the natural assets that underpin multiple uses of public lands.

---

[5] Text modified after the prepublication release to correct the allocation of funds to DOI and USDA for the National Seed Strategy from the Infrastructure Investment and Jobs Act: Public Law 117-58.

[6] Great American Outdoors Act: Public Law 116-152. Footnote added after the prepublication release for clarity.

[7] See https://www.doi.gov/priorities/america-the-beautiful (accessed February 14, 2023).

[8] See https://www.whitehouse.gov/ostp/news-updates/2022/04/24/accounting-for-nature-on-earth-day-2022/ (accessed February 14, 2023).

REPORT ORGANIZATION

A list of Acronyms and Abbreviations and a Glossary are included in the Front Matter of the report before the Summary. Chapter 2 provides an overview of the information-gathering strategy described in the interim report, including the methods used to identify target populations for surveys of state agencies and seed suppliers, and of semi-structured interviews of a sample of federal agency personnel.

Sitting at the end of the native seed supply chain are the users of native seed. They are a diverse group, with the federal government the largest. Chapter 3 focuses on the five agencies managing the most federal land. Chapter 4 explores the uses of native seed in state programs, and the activities surrounding that usage, through a survey of state staff discussing seed use, decision making, and concerns about the supply chain. Chapter 5 looks at native seed uses on tribal lands, along with a description of historical factors that affect tribal native seed capacities. Chapter 6 examines cooperative partnerships for native seed and plant development, some involving federal agencies, and others pursued at the state and regional levels. It also describes the effect of federal programs on the use of native seed on private lands. Seed suppliers, their activities, and their concerns are the focus of Chapter 7, with results from the survey of seed suppliers. Additional science is critical in support of restoration and seeds, and Chapter 8 discusses specific areas of basic and applied research that are needed to fully support the native seed supply. Chapter 9 provides important conclusions and overarching recommendations for actions to strengthen the native seed system.

The appendixes include committee biographies (Appendix 1); the federal agency interview tool (Appendix 2A); the state survey invitation letter (Appendix 2B), survey questions (Appendix 2C), and tabulated summary responses (Appendix 2D); the supplier invitation letter (Appendix 2E), survey questions (Appendix 2F), and tabulated summary responses (Appendix 2G); and the list of presentations made to the committee since the inception of the study over the years 2019–2022 (Appendix 2H).

CONCLUSIONS

Conclusion 1-1: *Native seed needs are defined by both the characteristics of the seed desired by users for ecological restoration and other purposes, and by the overall goal that requires their use.*

Conclusion 1-2: *In 2002, the Department of the Interior and Department of Agriculture put forward a plan for developing a native plant seed supply involving greater interagency coordination, partnerships with states and the private sector, and monitoring of restoration outcomes. Native seed shortages continue to be a barrier to restoration, and the aspirations of the plan, which is conceptually sound, have not been fulfilled.*

Conclusion 1-3: *The nation's native seed needs depend on the native plant communities that are the source of seeds. There is urgency to the need for conserving the biodiversity that is present in existing native plant communities and, therefore, for building a native seed supply through seed collection, plant development, and restoration, because climate change, extreme events, and destructive human activities have put these genetically and ecologically valuable natural resources at increasing risk.*

Conclusion 1-4: *Because more than half of the land in 11 western states is under federal management, it is likely that federal seed purchases, especially after wildfires, have a major influence on the native seed industry in the western United States.*

Conclusion 1-5: *Recent congressional legislation and executive orders have provided an unprecedented opportunity to address the nation's natural heritage on public lands.*

REFERENCES

Barga, S.C., P. Olwell, F. Edwards, L. Prescott, and E.A. Leger. 2020. Seeds of Success: A conservation and restoration investment in the future of US lands. Conservation Science and Practice 2(7):e209.

De Vos, J.M., L.N. Joppa, J.L. Gittleman, P.R. Stephens, and S.L. Pimm. 2015. Estimating the normal background rate of species extinction. Conservation Biology 29:452–462. https://doi.org/10.1111/cobi.12380.

Havens, K., P. Vitt, S. Still, A.T. Kramer, J.B. Fant, and K. Schatz. 2015. Seed sourcing for restoration in an era of climate change. Natural Areas Journal 35(1):122–133.

NASEM (National Academies of Sciences, Engineering, and Medicine). 2020. An Assessment of Native Seeds and the Capacity for Their Supply: Interim Report. Washington, DC: The National Academies Press. https://doi.org/10.17226/25859.

NICC (National Interagency Coordination Center). 2021. Wildland Fire Summary and Statistics Annual Report: 2020. https://www.predictiveservices.nifc.gov/intelligence/2020_statssumm/intro_summary20.pdf. (accessed February 9, 2023).

NOAA (National Oceanographic and Atmospheric Administration). 2021. Record-breaking Atlantic hurricane season draws to an end. https://www.noaa.gov/media-release/record-breaking-atlantic-hurricane-season-draws-to-end (accessed February 9, 2023).

PCA (Plant Conservation Alliance). 2015. National Seed Strategy for Rehabilitation and Restoration. www.blm.gov/seedstrategy (accessed March 31, 2020).

Pimm, S.L., C.N. Jenkins, R. Abell, T.M. Brooks, L. Gittleman, N. Joppa, P.H. Raven, C.M. Roberts, and J.O. Sexton. 2014. The biodiversity of species and their rates of extinction, distribution, and protection. Science 344(6187):1246752. https://www.science.org/doi/10.1126/science.1246752.

Scheffers, B.R., L. De Meester, T.C. Bridge, A.A. Hoffmann, J.M. Pandolfi, R.T. Corlett, S.H. Butchart, P. Pearce-Kelly, K.M. Kovacs, D. Dudgeon, M. Pacifici, C. Rondinini, W.B. Foden, T.G. Martin, C. Mora, D. Bickford, and J.E. Watson. 2016. The broad footprint of climate change from genes to biomes to people. Science 354(6313):aaf7671. doi:10.1126/science.aaf7671.

Vincent, C.H., L.A. Hanson, and L.F. Bermejo. 2020. Federal Land Ownership: Overview and Data, Updated February 21, 2020. Congressional Research Service Report R42346. https://crsreports.congress.gov/ (accessed February 1, 2023).

Williams, A.P., B.I. Cook, and J.E. Smerdon. 2022. Rapid intensification of the emerging southwestern North American megadrought in 2020–2021. Nature Climate Change 12(3):232–234.

2

Description of the Information-Gathering Strategy

The Committee undertook several information-gathering activities to obtain a thorough understanding of the native seed supply chain. The data-gathering strategy varied, depending on the specific type, projected use, and availability of the information. This section describes the main activities, which included (a) semi-structured interviews with federal seed buyers, (b) surveys of state government seed buyers, (c) surveys of seed suppliers, (d) public information-gathering sessions, (e) a review of existing data, and (f) a review of the published literature.

SEMI-STRUCTURED INTERVIEWS WITH FEDERAL SEED BUYERS

To better understand native seed purchases at the federal government level, the Committee conducted a series of semi-structured interviews with staff from five federal agencies, including the Bureau of Land Management (BLM), the US Forest Service, the National Park Service, the US Fish and Wildlife Service, and the military service branches (the Army, Air Force, Navy, and Marine Corps). Within each agency, the Committee attempted to identify a few key staff members who were especially likely to have broad knowledge about the use of native seed and plant materials within the agency.

Appendix 2A shows the interview guide, which was developed by the Committee and covered the following broad topics:

- Purchase and use of native seed and plant materials
- Contracting arrangements
- Sources of information about native seed availability
- Communication with suppliers
- Decision making about the use of native seed and plant materials
- Substitutions when the desired natives are not available
- Monitoring of projects after planting
- Expectations for the future

The interviews were conducted by the study director (the director of the Board on Agriculture and Natural Resources) with assistance from a research associate. A total of 19 interviews were completed between August of 2021 and May of 2022. The interviews were conducted via Zoom or telephone and were typically 1–1.5 hours in length.

SURVEYS

The Committee conducted two surveys, one of seed suppliers and another one of state government agencies that buy native seed. This section describes the methods used to collect the survey data.

Survey of State Government Seed Buyers

The Committee conducted a survey of departments within state government that use seed and plant materials. Appendix 2C shows the wording of each of the questions included in the survey, which covered the following broad topics:

- Purchase and use of native seed and plant materials
- Sources of native seed and plant materials
- Importance of specific seed characteristics
- Purposes of native seed purchases
- Seed storage availability
- Contracting arrangements used
- Sources of information about native seed availability
- Communication with suppliers
- Substitutions when the desired natives are not available
- Monitoring of projects after planting
- Availability of in-house expertise
- Barriers to using native seed
- Expectations for the future

The survey was carried out by the Social and Economic Sciences Research Center at Washington State University (SESRC) based on specifications provided by the committee. SESRC staff also provided input on the survey questions and the implementation methods.

The survey included state agencies that are most likely to use seed and plant materials due to their mission. The list of departments was compiled by National Academies of Sciences, Engineering, and Medicine staff and included departments focused on the general areas of natural resources, forestry, fish and wildlife, state parks, and transportation. States vary in how the relevant departments are structured and named, and staff used public websites to generate the list, along with contact information for the heads of each department (such as director or commissioner). The final list contained 160 departments (an average of 3.2 departments per state). The survey was sent to all 160 departments. While the goal was to include all departments that are likely to use native seed and plant materials, the list was not a comprehensive list of all state departments that could potentially do so. These results should not be generalized to all state departments in the United States beyond the list of departments compiled for this survey. The survey included departments of different sizes from all regions of the country (see below).

The survey was carried out between May and July of 2021, online and by telephone. The survey was conducted during the COVID-19 pandemic, when many state agency employees were working from home, so the first contact was an email sent to the department head. The letter provided a link and access code for the online survey and asked that the survey be completed by the person who is most knowledgeable about the use of native seed and plant materials in the department (see Appendix 2B). Reminders were sent by email, and trained SESRC interviewers also conducted telephone follow-up in the final stages of the fieldwork for those who did not complete the survey online. A small subset (3%) of the departments that responded completed the survey by telephone.

Table 2-1 shows the final outcome of the cases included in the survey, including the cases that were determined to be ineligible because the department did not use either native seed or plant materials. See Appendix 2C for the screening questions that were used to determine eligibility. The response rate was 63 percent.

The survey included both closed and open-ended questions. Open-ended questions were coded into thematic categories by two trained SESRC coders. For each question, the coders created a codebook together with one coder starting the codebook and the other testing the validity of each code and further refining the codebook as necessary.

TABLE 2-1 Outcome of Cases for the Survey of Departments within State Government Agencies

Case Disposition	Frequency	Percent
(A) Completed	92	57
(B) Partially completed	5	3
(C) Refusal to participate	3	2
(D) Non-response	52	33
Total eligible	**152**	**95**
Response Rate (A+B)/(A+B+C+D)	**Response Rate: 63%**	
(E) Ineligible (not using native seed or plant materials)	8	5
Total ineligible	**8**	**5**
Total sample	**160**	**100**

The coders regularly met throughout the coding process to compare the cases coded and discuss discrepancies. After completing the coding for each question, 20 percent of cases were randomly selected for an intercoder reliability check. The coders then used the codebook to code those cases again to ensure at least 80 percent agreement on all codes. While all codes were above the 80 percent threshold, the coders discussed discrepancies until agreement was found and revised the codebook where necessary.

Table 2-2 shows the distribution of the responses by geographic area. While the geographic area categorized as the East for the purposes of this report includes a larger number of states and therefore a larger number of departments than the West, the response rate was higher among the Western states than the East. It is possible that more departments from the eastern states felt that the survey did not apply to them and chose not to respond. Table 2-3 shows the distribution of the responses by size of the department's expenditures on natives, determined based on the approximate average annual expenditure on native seed and plant materials between 2017 and 2019. Key results from the survey, including the open-ended questions, are discussed in subsequent chapters of this report. Response frequency distributions for the questions are shown in Appendix 2D.

Survey of Seed Suppliers

To better understand the experiences of seed suppliers, the Committee conducted a survey of sellers of native seed and plant materials. Appendix 2F shows the wording of each of the questions included in the survey, which covered the following broad topics:

- Types of business activities
- Types of seed and plant materials sold
- Capacity for collecting seed
- Capacity for growing seed
- Seed storage availability

TABLE 2-2 Distribution of Responses to the Survey of State Government Departments by Geographic Region

Region	Number of Departments	Percent
East	65	67
West	32	33
Total	97	100

NOTE: East includes AL, AR, CT, DE, FL, GA, IA, IL, IN, KS, KY, LA, MA, MD, ME, MI, MN, MO, MS, NC, ND, NE, NH, NJ, NY, OH, OK, PA, RI, SC, SD, TN, TX, VA, VT, WI, WV; West includes AK, AZ, CA, CO, HI, ID, MT, NM, NV, OR, UT, WA, WY.

TABLE 2-3 Distribution of Responses to the Survey of State Government Departments by Size

Department Expenditures on Natives	Number of Departments	Percent
$100,000 and below	43	56
Over $100,000	26	34
Don't know	8	10
Total responses to question	77	100
Missing	20	
Total	97	

NOTE: Size is based on the question: "Thinking about the three-year period from 2017 to 2019, what was [DEPARTMENT]'s approximate average annual expenditure on native seed and plant materials combined?"

- Types of buyers and methods of communication with potential buyers
- Contracting arrangements used
- Challenges and barriers encountered
- Ability to plan ahead
- Expectations for the future
- Suggestions for improving how the market for native seed functions

This survey was also carried out by SESRC based on specifications provided by the committee. SESRC staff provided input on the survey questions and the implementation methods.

Given that there is no definitive list of all seed suppliers that could have served as a sampling frame for the survey, the Committee built a list based on information from the following sources that contained the names and contact information of vendors with an interest in native seed:

- US Department of Agriculture National Nursery and Seed Directory (Native Plants Journal list)
- Native Seed Network
- American Seed Trade Association's Environmental and Conservation Seed Committee
- BLM seed vendors
- US Forest Service seed vendors

After a list based on the above sources was assembled, the list was deduplicated, and entities that were not commercial vendors located in the United States were removed. The deduplication was primarily focused on identifying duplicate organization names, but the list was small enough that staff were able to also manually flag and resolve a few additional exceptions, such as two separate organization names listed with the same address (due for example to name changes or acquisitions). The final list contained 1,259 suppliers, and the survey was a census of all the suppliers on the list (in other words, not a sample survey of a subset of the suppliers). As discussed, this was not a comprehensive list of all suppliers of native seed, so the results cannot be generalized to all suppliers in the United States. However, the survey included the suppliers used by BLM in recent years. The survey also included non-BLM suppliers of different sizes from all regions of the country (see below), but it is ultimately not possible to know how this list compares to the theoretical universe of all native seed suppliers in the country.

The survey was carried out between May and July of 2021, online and by telephone. The first form of contact with those selected into the survey was a physical letter mailed to the business. The letter provided a link and access code for the online survey and asked that the survey be completed by the person who is most knowledgeable about the sales of native seed and plant materials at the particular business (see Appendix 2E). The letter was followed by both email and postal reminders. Trained SESRC interviewers also conducted telephone follow-up in the final stages of the fieldwork for those who did not complete the survey online. This approach resulted in 78 percent of the completed interviews being completed online and 22 percent by telephone.

TABLE 2-4 Outcome of Cases for the Supplier Survey

Case Disposition	Frequency	Percent
(A) Completed	223	18
(B) Partially competed	48	4
(C) Refusal	83	7
(D) Non-response	650	51
Total eligible	**1,004**	**80**
Response Rate (A+B)/(A+B+C+D)	**Response Rate: 27%**	
(E) Doesn't sell native seed or plants	78	6
(F) Sells only to the home market	110	9
(G) No longer in business	65	5
(H) Other	2	0
Total ineligible	**255**	**20**
Total sample	**1,259**	**100**

Table 2-4 shows the final outcome of the cases included in the survey. The response rate, defined as the ratio of completed and partially completed interviews to the total eligible cases, was 27 percent. Approximately 20 percent of the cases were determined to be ineligible on the basis of their responses to a set of screening questions (see Appendix 2F), and in some cases information obtained from other sources. Ineligible cases included vendors that were no longer in business or were not selling native seeds or plants. Due to the focus of this study, vendors that were exclusively focused on the home market were also excluded from the survey.

The survey included both closed and open-ended questions. Open-ended questions were coded into thematic categories by two trained SESRC coders. For each question, the coders created a codebook together with one coder starting the codebook and the other testing the validity of each code and further refining the codebook as necessary. The coders regularly met throughout the coding process to compare the cases coded and discuss discrepancies. After completing the coding for each question, 20 percent of cases were randomly selected for an intercoder reliability check. The coders then used the codebook to code those cases again to ensure at least 80 percent agreement on all codes. While all codes were above the 80 percent threshold, the coders discussed discrepancies until agreement was found and revised the codebook where necessary.

Table 2-5 shows the distribution of the responses by geographic area and Table 2-6 shows the distribution of the responses by the size of the business, determined based on the average annual sales and operating revenues between 2017 and 2019. As discussed, an inventory of all native seed suppliers and their characteristics does not exist, so it is not possible to determine how these breakdowns compare to the universe of vendors as a whole. Key results from the survey, including the open-ended questions, are discussed in subsequent chapters of this report. Response frequency distributions for the questions are shown in Appendix 2G.

TABLE 2-5 Distribution of Responses to the Supplier Survey by Geographic Region

Region	Number of Suppliers	Percent
East	147	54
West	124	46
Total	271	100

NOTE: East includes AL, AR, CT, DE, FL, GA, IA, IL, IN, KS, KY, LA, MA, MD, ME, MI, MN, MO, MS, NC, ND, NE, NH, NJ, NY, OH, OK, PA, RI, SC, SD, TN, TX, VA, VT, WI, WV; West includes AK, AZ, CA, CO, HI, ID, MT, NM, NV, OR, UT, WA, WY.

TABLE 2-6 Distribution of Responses to the Supplier Survey by Supplier Size

Supplier Size	Number of Suppliers	Percent
$499,999 or less	112	53
Between $500,000 and $4,999,999	67	32
$5,000,000 or more	30	14
Don't know	1	1
Total of respondents to question	210	100
Missing	61	
Total	271	

PUBLIC INFORMATION-GATHERING SESSIONS AND OTHER PUBLIC INPUT

In addition to the structured data-collection efforts, an important source of input for the Committee was the public information-gathering meetings held with a variety of stakeholders and experts working in areas related to native seed and plant materials. The Committee met with representatives from BLM, and with staff from other federal and state government agencies that have projects related to native seed. The Committee also met with suppliers of native seed, operating businesses of various sizes. Other stakeholders who provided input to the Committee included representatives of organizations with an interest in native seed and plant materials, researchers, seed cleaners, seed certifiers, and seed storage providers. Appendix 2H shows the list of individuals who participated in meetings with the committee. In addition, a call for public comments was also posted to the study page on the National Academies website. The website provided a link to a form for submitting comments to the Committee.

REVIEW OF EXISTING DATA

To learn as much as possible about the native seed supply chain, the committee obtained relevant information that was available in the form of administrative records held by various agencies and other entities that are part of the supply chain. While the committee was not aware of any existing data source that could provide a comprehensive picture of either the native seed supply or demand in the United States, these records provided a granular view of some segments of the supply chain and enriched the committee's understanding of a few specific market players. The main sources of existing data reviewed by the committee included consolidated seed buy records from BLM.

REVIEW OF PUBLISHED LITERATURE

The Committee's work builds on a rich literature that exists on a broad range of topics related to the role and use of native seed and plant materials. The Committee reviewed recent research, with particular focus on topics related to the study charge. These topics ranged from restoration successes to the challenges associated with the use of native seed, as well as research on choosing native seed. The Committee also benefited from the published literature on the experiences of other countries and groups focused on specific geographic areas, and from gaining an understanding of a variety of approaches, systems, and processes.

3

Native Seed Needs and the Federal Government

The first section of this chapter provides a high-level view of what is driving native seed needs of the federal land-management agencies, the scale of the need, and how each agency is attempting to address those needs. The second part provides a summary of collective insights from individuals working in those agencies to shed light, from a boots-on-the-ground perspective, of the day-to-day challenges related to native seed used in ecological restoration and other purposes. The information in the first part is derived from public presentations by agency representatives and public information about agency activities. The second part is based on semi-structured interviews with agency personnel. The chapter ends with a discussion of the significance of the findings and the committee's conclusions.

TOP FIVE FEDERAL LAND-MANAGEMENT AGENCIES

The Committee's Interim Report identified the five largest federal land-management agencies with management responsibilities for more than 600 million acres (Table 3-1), the amount of land each manages, and provided a brief description of the mission of each agency. Each of these agencies makes direct purchases of native seed. The four largest agencies have a major environmental focus as part of their missions.

Bureau of Land Management

As its name implies, the Bureau of Land Management (BLM; within the Department of the Interior) is the federal government's largest land manager and uses the largest quantities of native plant seed in the nation. Thus, it is particularly important to the Committee's work. BLM manages extensive regions in the western states, and oversees approximately a tenth of the overall land area in the United States. Its mission directly relates to sustaining and utilizing natural resources on that land.

BLM's Plant Conservation and Restoration Program (PCRP) operates out of the Bureau's Resources and Planning Directorate to ensure that seeds are available for the projects of over 100 BLM field offices in 11 western states. Most of the seed purchased by BLM is through Consolidated Seed Buys which typically take place three or four times a year.[1] The procurement process begins with the development of a list of seed types and amounts requested from the field offices, which is then put out for public bid. The winning bids are paid from a Working

---

[1] Text was modified after the prepublication was released to correct number of BLM offices with seeding projects and the number of Consolidated Seed Buys annually.

TABLE 3-1 Land Managed by Five Major Federal Agencies (acres, 2018)

Bureau of Land Management (BLM)	244,391,312
US Forest Service (USFS)	192,919,130
US Fish and Wildlife Service (USFWS)	89,205,999
National Park Service (NPS)	79,045,679
Department of Defense (DOD)	8,845,476
Total for Five Agencies	615,311,596

SOURCE: US Congressional Research Service, 2020.

Capital Fund, which is then reimbursed from field offices or other project-specific sources. The purchased seed is stored temporarily in one of BLM's seed warehouses until used.

Native Seed and Wildfire

Wildland fire is the major driver for native seed needs for BLM, and for other federal land-management agencies, as well as for all state and tribal land in the West. As Table 3-2 shows, wildfire is a frequent occurrence and has significant impacts on all public and private lands in the West, with more than 9.5 million acres burned in 2020, and over 1.1 million on BLM land.

After wildfires, seeds are used to establish vegetation to stabilize and protect soils and assist the process of natural recovery. Because the number and severity of wildland fires vary from year to year, the total annual seed purchases from the Consolidated Seed Buys fluctuates annually, for example, from a high of 7.5 million pounds in 2007 to a low of approximately 300,000 pounds in 2009. BLM estimates that on average, 2.4 million pounds of pure

TABLE 3-2 Wildfires in Western United States in 2020 (number and acres across all jurisdictions, and the acres under BLM management)

	All Lands		BLM Lands
	Number	*Acres*	*Acres*
Alaska	349	181,169	45,256
Arizona	2,524	978,568	76,204
California	10,431	4,092,150	142,201
Colorado	1,080	625,357	167,723
Idaho	944	314,352	60,147
Montana	2,433	369,633	11,670
Nevada	770	259,276	223,052
New Mexico	1,018	109,513	6,809
Oregon	2,215	1,141,613	234,047
Utah	1,493	329,735	102,355
Washington	1,646	842,370	30,180
Wyoming	828	339,783	30,390
Total	**25,731**	**9,583,519**	**1,130,034**

SOURCE: Public Land Statistics 2020, Department of the Interior, Bureau of Land Management. Available at https://www.blm.gov/sites/default/files/docs/2021-08/PublicLandStatistics2020_1.pdf (accessed February 9, 2023).

live seed are purchased each year through this process, at an average annual cost of $20 million. About 30–40 seed suppliers each year respond to the Consolidated Seed Buys (Ricardo Galvin, presentation to the committee).

In a presentation to the committee, Molly Anthony, the Program Lead for the BLM Emergency Stabilization and Rehabilitation (ESR) program noted that in 2020, the field offices of the ESR program purchased $14.9 million in seed and conducted post-fire seeding on over 422,000 acres of burnt land. Anthony explained to the committee the institutional limitations on the effectiveness of post-fire emergency stabilization and rehabilitation, which accounts for fully two-thirds of BLM seed procurement. Seed purchases for this program are dependent on funds allocated for the current fiscal year, not future years. Further, federal spending rules curtail purchasing for the current year in a black-out period before September 30, the end of the federal fiscal year, even though most seeding is conducted in fall. These institutional restrictions limit the capacity for advance planning and procurement and make the program dependent on seed supplies that are already in warehouses or commercially available. In turn, desired seeds are often unavailable, and substitutions of less-desired seeds are common.

Seeds purchased through the Consolidated Seed Buys are grasses, forbs (small flowering plants that are not grasses), and shrubs. The major portion of seed purchases is grasses, of which more than half (by weight) of seed requests from field offices have been for native species. In contrast, the large majority (by weight) of forb seed purchases are of non-native species. Field office requests for native forbs have exceeded what is provided in the purchases. Purchases of shrub seed (mainly related to sagebrush) are about two-thirds native (Patricia Roller, personal communication, based on historical BLM seed purchase data).

Seed Transfer Zones and Source-Identified Seed

BLM does not have an agency-wide requirement that seed purchased by the field offices be native plant seed, although its Best Management Practices call for the use native seed of known origin when available (BLM, 2008)[2] and its use is encouraged by program offices, such as the ESR unit. The PCRP would prefer that field offices request and buy native, certified Source-Identified (SI) seed, for which the geographic location of the origins of the seed was verified by an official seed certification agency. Having that information increases the likelihood of selecting a seed type from a location or "seed transfer zone"[3] that is appropriate for where it will be used, using guidance developed by the US Geological Survey, US Forest Service (USFS), and other research agencies that have studied the genetics of plant species across the landscape.

However, it is unlikely that the field offices will or can meet this preference. In 2020, BLM field units purchased 1.4 million pounds of grasses, 220,000 pounds of forbs, and 110,000 pounds of shrub seed. Of these seeds, about one-eighth of the grasses, one-fourth of the forbs, and two-thirds of the shrubs were native SI seed (BLM Seed Purchase data). The rest of the seed was either non-native seed or seed of released native germplasms or cultivars.[4]

It is quite possible that the type of seed desired by the field offices is simply not available from suppliers, so buyers have had to accept substitutions, as Anthony described. However, the choices made between different types of seed (native vs. non-native, grass vs. forb, cultivar vs. source-identified) also appear correlated with the costs per pound of these types, so price is likely to be a factor in the decision making of BLM field offices.

Proactive Stockpiling of Native, Source-Identified Seed with the IDIQ

The PCRP has no authority to require that field offices purchase native seed or native SI seed, and until recently, has had no budget with which to stock BLM warehouses with native, SI seed in advance of post-fire operations and other uses. However, in 2018, BLM created a seed production component of its Working Capital Fund and established an Indefinite Delivery Indefinite Quantity (IDIQ) procurement vehicle designed to streamline

---

[2] The text was modified after the prepublication release to clarify the BLM policy on the use of native plant seed.

[3] Seed transfer zones are geographically distinct areas within which seeds can be sourced and introduced with low risk of maladaptation (Bower et al., 2014; Breed et al., 2013), and minimal loss of biodiversity (Malaval et al., 2010).

[4] Most of the native releases are genetically non-manipulated (not created through selection or breeding), but due to their greater availability and less expensive pricing in the marketplace, they are often planted in geographic areas not matching their geographic source; that is, they originate from a different seed transfer zone than the one where they will be used.

contracting for seed increases.[5] As a result, contracts were awarded to produce SI seed collected by the BLM Seeds of Success program (discussed in Chapter 6), with a focus on native grasses and forbs from six ecoregions.[6]

The goal of the program is for the seed increases to be completed over 3–5 years, and then stored in the BLM warehouses for future purchase by field offices when needed. Currently, there are six suppliers holding IDIQ contracts and the award has a 5-year cap at $49 million total, significantly less than the approximately $20 million average each year spent on the Consolidated Seed Purchases.

It is a significant first step toward the proactive stockpiling of native seed to meet the Bureau's needs. Between 2019 and 2021, purchases through the IDIQ included 42 species and 94 combinations of taxa and seed zones, of which approximately 94,200 pounds of seed are expected to be delivered in the next 3 years (PCA, 2022). Anne Halford, BLM Idaho State Botanist, told the committee that the IDIQ contract structure is meant to increase the supply of priority native forbs and so-called workhorse species (widespread, quickly establishing, native plant species) for targeted seed zones. She stated that about 85 percent of BLM's seeds are used in three[7] seed zones in the Great Basin. That fact, she noted, underscores the importance of improving seed source protection of the wild plant populations that exist in these seed zones, and for collection of seeds from these areas for seed banking and future increase.

Warehouse Needs

The IDIQ initiative is promising, though it is unclear at this stage how far the initiative can be expanded to meet the Bureau's growing needs. In addition, if BLM were able to upscale native seed and plant material supplies, the current National Seed Warehouse System would soon be inadequate in terms of physical climate-controlled capacity, staff, and expertise. Currently, BLM has two major warehouses, one staffed, state seed warehouse in Boise, Idaho, and the other in Ely, Nevada, which combined can accommodate 2.6 million pounds of seed (Patricia Roller, BLM, presentation to the committee). There is some refrigerated storage space at the Ely site. The Ely warehouse was not staffed earlier in 2022 (BLM, personal communication).

Other Needs for Native Seed Within BLM

Although post-wildfire stabilization and long-term rehabilitation purchases the most seed through the BLM seed procurement processes, there are several other important units that also need seed to fulfill their targeted missions in BLM, including accompanying fuels reduction; wildlife habitat, including pollinators; forest and riparian areas; rangeland; recreation; mines; oil and gas wells; and other energy development.

In a presentation to the committee, Anne Halford, BLM Botanist in Idaho, emphasized the different ways in which plant conservation and restoration activities contribute to revenue generation for the Department of the Interior and BLM, from the $23 million in recreation fees in 2017 that brought visitors to witness the "super bloom" of native wildflowers, to supporting the preferred native habitat for fish and game species on which $348 million is spent annually, and the increased productivity of rangeland on which ranchers in the West rely.

An important consideration related to the current biodiversity crisis is the ecosystem value of native plant communities as food and shelter for the iconic sage-grouse or other rare species. Plant materials are also needed to address orphaned oil and gas wells on federal lands, for which BLM has taken the lead with remediation funds of $250 million in the Bipartisan Infrastructure Law.[8] There are massive numbers of orphaned wells in the country

[5] The text was modified after the prepublication release to clarify that the Working Capital Fund can support the IDIQ but the two are not intrinsically linked.

[6] The Environmental Protection Agency (Omernik, 1987) created a map of ecosystems (ecoregions) across the United States defined by major differences in geography and climate. Subsequent refinements of the map have led to more discrete ecoregions.

[7] Development of a climate-smart restoration tool led researchers to suggest combining provisional seed zones for sagebrush in the Great Basin into three zones, making it easier to source seed appropriate for the Great Basin region.

[8] Infrastructure Investment and Jobs Act (Bipartisan Infrastructure Law), see https://www.blm.gov/sites/default/files/docs/2021-05/Congressional_20210415_HR2415.pdf (accessed February 9, 2023).

FIGURE 3-1 Oil and gas well pads in Jonah Field, Wyoming. SOURCE: Ecoflight.

(over 130,000 by one estimate[9]) that dot the national landscape (see Figure 3-1). The Bipartisan Infrastructure Law (BIL) includes $4.4 billion for the Department of the Interior to provide grants to the tribes, states, and private landowners to plug the wells and restore the soil and habitat, at an average cost of $25,000 per well pad. One study estimated that to address the 1 to 4 percent of west Texas land with well pads and pipelines between now and 2050 would require between 247,000 and 1,330,000 pounds of seed of native grasses valued between $10 million and $57 million (Smith et al., 2020).

US Forest Service

The mission of the USFS (within the US Department of Agriculture) is to sustainably manage national forest land for multiple uses. The agency also supports management of forest land under private, state, and tribal ownership. Trees species are the major focus of the agency's seed collection, development, and seedling production activities, but in the last decade that has expanded to include grasses and forbs.

Wildfire and Restoration Needs

Like BLM, wildfire is driving the restoration needs of the USFS. In 2020, the USFS Pacific Northwest Region 6, home to 17 National Forests across Oregon and Washington, experienced the worst fire season in a century. According to a Rapid Assessment Team[10] report evaluating the Slater Fire in the Rogue River-Siskiyou

---

[9] See https://iogcc.ok.gov/sites/g/files/gmc836/f/documents/2022/iogcc_idle_and_orphan_wells_2021_final_web_0.pdf (accessed February 9, 2023).

[10] See https://www.fs.usda.gov/Internet/FSE_DOCUMENTS/fseprd887614.pdf (accessed February 9, 2023).

BOX 3-1
A New Level of Forest Management on Public Lands

In January of 2022, the USFS released Confronting the Wildfire Crisis Strategy, calling for a "paradigm shift" in the scale and pace at which fuels removal and forest health treatments take place, anticipating four times as much intensive management activity than in the past. The 10-year plan will treat 20 million acres of USFS property and 30 million acres of other federal, state, tribal and private land.

According to the Strategy, the goal is to manage the lands in the West as one landscape, with the work carried out under the direction of the National Interagency Fire Center in Boise, Idaho, where agencies have worked together "seamlessly" on wildfire since 1965.

The large-scale effort to remove dead trees, conduct thinning, and other work will be complemented by a National Forest System Reforestation Strategy, with funds from the REPLANT (Repairing Existing Public Land by Adding Necessary Trees) Act, part of the Infrastructure Investment and Jobs Act known informally as the Bipartisan Infrastructure Law (BIL) passed in 2021. The Act will result in as much as $123 million annually available for reforestation activities from import duties on wood products, which will support the reforestation effort of 4.1 million acres over the next 10 years.

National Forest, a half million acres of USFS-managed land was burned by wildfires in 2020. Seventy-five percent of the trees in the core fire area (about 200,000 acres) were killed. The report said the event would "more than double our existing reforestation needs in the region and will put a strain on seed or seedling availability for some forests." The report added that the USFS Region 6 Geneticist (Vicky Erickson) had already reached out to develop agreements with other land-management agencies from which seed might be used or purchased and was working with other USFS geneticists to identify genetically appropriate tree seed sources that might be acquired.

The pressing need for an adequate supply of tree seeds for restoration after wildfire is reminiscent of the challenges that BLM faces annually. The USFS has generally preferred to allow burned forest to recover naturally and has previously lacked the resources to address its 4-million-acre backlog in reforestation needs—the USFS estimates that only 6 percent of current reforestation needs are met annually.[11] In 2017, the agency reported that more than half of its budget (including non-fire programs) was used to fight wildfires.[12] In the wake of two extreme fire years, in January 2022, the USFS released *Confronting the Wildfire Crisis Strategy*, a 10-year plan to dramatically increase fuels removal and forest health activities on millions of forested acres, which was followed in July 2022 by the *National Forest System Reforestation Strategy*, a plan to address the reforestation deficit (see Box 3-1).

Seed Acquisition through the BPA

To address seed needs, Region 6 has in-house capacity for tree production at the Clarno Propagation Center, and production of conifer trees, grasses, and other plant types at the J. Herbert Stone Nursery. Because many more trees and other native plants will be needed, production and other activities will need to be scaled up to engage many more public and private nurseries and growers.[13] In a presentation to the committee, Erickson said Region 6 uses a Blanket Purchase Agreement (BPA) to commission this work, having switched from an IDIQ in 2017. The advantages of the BPA over the IDIQ it used previously included no up-front obligations, a higher dollar spending limit ($10 million), and the ability to add new vendors over the 10-year term of the agreement. She added that the USFS BPA is available to other federal agencies, and overall provides a relatively fast and easy mechanism for accessing vetted contractors and quickly obligating funds as they become available.

---

[11] See https://www.usda.gov/sites/default/files/documents/reforestation-strategy.pdf (accessed February 9, 2023).

[12] See https://www.fs.usda.gov/sites/default/files/fy-2017-fs-budget-overview.pdf (accessed February 9, 2023).

[13] See https://www.americanforests.org/wp-content/uploads/2021/02/Ramping-Up-Reforestation_FINAL.pdf (accessed February 9, 2023).

Work orders under the BPA list grow-out specifications for each species. The pool of qualified growers can submit proposals on a one-page form. Seed stock is provided by the agency, either collected on USFS land by agency staff or via contracting, following guidelines for obtaining genetic diversity and recording of the collection site data. Growers are normally paid per pound, but per-acre prices can be used for higher-risk species or small quantities. Production data are maintained to guide future seed increase. Compared with previous contract structures, Erickson said the BPA is better in recruiting good producers (though there are still not enough of them), being flexible and fast, reducing risk to producers, maintaining quality control, and building expertise. The USFS uses its BPA to contract for a wide range of services, such as invasive plant treatment, cone collection, monitoring, and other tasks.[14]

Erickson noted that an important goal is for seed production to benefit rural economies, both through long-term partnerships for the USFS and through secondary markets (commercial seed sales). For example, the White River National Forest in Colorado initiated a native grass seed development program through collections and increases that ultimately resulted in the release of certified SI seed of two grass species (slender wheatgrass and mountain brome) to commercial suppliers in Colorado, which are now able to grow these grass species in large quantities to sell to private landowners and to land-management agencies.

The USFS has an institutional policy directing the National Forests to cooperate in developing and using native plants. However, each National Forest has decision-making autonomy and its own priorities. Forests vary in their approach to native seeds. Region 6 has a national reputation for a well-functioning native plant and genetics program that plays a strong advisory role to the National Forests in its region, which includes developing projections of seed needs in the National Forests as the climate changes, assisting in the collection of seed from the National Forests, and the processing, storage, and planting of seeds and seedlings. The USFS does not have large seed warehouses like BLM, but in Region 6 there is some ambient storage at the Bend Seed Extractory (in Bend, Oregon), which is a seed cleaning and conditioning facility that extracts conifer seeds from their cones and cleans the smaller, tiny seeds of more than 3,500 other native plant species prior to seed viability testing. The Bend facility is available to clean seed for other public agencies and is the seed cleaning facility of choice for the BLM Seeds of Success (SOS) program.

US Fish and Wildlife Service

The USFWS (within the US Department of the Interior) currently holds title to the chair position of the Federal Committee of the Plant Conservation Alliance and participates in many activities that contribute to the fulfillment of the National Seed Strategy. The USFWS administers the National Wildlife Refuge System, consisting of more than 560 wildlife refuges and administers the Endangered Species Act of 1973. A key regulatory and advisory function of the USFWS is carried out through consultations with public and private entities to moderate activities that might affect threatened and endangered species (TES) of plants and animals.

The Service uses native seeds for conservation activities, such as developing pollinator habitat, and conducting post-wildfire restoration, and invasive species removal in the refuges. The Service frequently works in partnership with other public and private organizations to carry out its activities related to protecting habitat. It uses voluntary conservation easements on private lands and other tools to assemble conservation areas to unite fragmented habitat and can acquire land and easements through the Land and Conservation Fund, which was recently bolstered with an annual appropriation of $900 million through the Great American Outdoors Act. In addition to wildlife, many native plants are themselves TES. There are 876 plants on the federal TES list that the Service seeks to conserve.

Wildfire and Climate Change Threats

Unlike the BLM and USFS, the National Wildlife Refuges are not concentrated in the West but spread across the entirety of the Unites States. More wildfires occur each year in the Midwest and the eastern United States (35,000 in 2021) but at much smaller scale (1.0 million acres) than in the West (23,000 fires, 6.2 million acres) (CRS, 2022).

---

[14] This section was modified to clarify that the USFS BPA is being compared to its previously used IDIQ, not the BLM IDIQ.

The USFWS reports that 398,000 acres of land under its management burn annually on average, but also notes that a large proportion of the wildlife refuges are fire adapted, that is, their ecosystems have evolved along with periodic, moderate fires that play a role in habitat regeneration and maintaining a balanced ecology. The Service conducts prescribed fires on many refuges as a fire management tool. It also carries out post-fire, Burned Area Rehabilitation (BAR) projects on the refuges in partnership with other organizations and volunteers that assist in the production and installation of native plants, seeding, and other rehabilitation activities. One example is the current activity on the Keālia Pond National Wildlife Refuge following the 2019 central Maui fire, in which the USFWS has commissioned a local nursery to produce "1000 native plants per month for the next three years" to be installed by Americorp volunteers on a weekly basis.

Just as wildfire is challenging, so are the daunting and complex impacts of climate change to the National Wildlife Refuges, given the variation and degrees of impact of climate change that will be unique to each refuge. The Service is adopting a Resist-Accept-Transition framework to help it chart a management course, that among other decisions, will affect its habitat restoration and TES decisions, both with implications for native plant seed and materials.

USFWS Partnerships in Support of Native Seed

The total amount of seed and plant materials procured across all the National Wildlife Refuges used for BAR projects or other, non-fire-related restoration projects is unknown, in part because the activities are widely distributed. The Department of the Interior (DOI) has received $325 million over 5 years (2021–2026) to expand BAR activities on lands of all the DOI federal land-management agencies.

The USFWS does not have its own seed production or storage facilities but does contract for seed collection on the National Refuges, uses the seed banking capacity of botanical gardens, such as through the Center for Plant Conservation, and collaborates with different nurseries, depending on the project, such as the Southern Highlands Reserve, for the collection and propagation of red spruce.

In 2022, the USFWS announced that $2 million in funds from the 2021 BIL will help it launch collaborative activities for the collection of native seed following SOS protocols. Notably, in partnership with the Southeastern Grasslands Initiative, the Service will collect and bank seed from across 10 states in the southeastern United States in a new SOS-Southeast program.

National Park Service

The NPS (within the Department of the Interior) manages over 400 individual parks, monuments, and other units with the mission of conserving these natural and historic lands and their wildlife to make these available for future generations. According to NPS Management Policy, Chapter 4, the Parks "will try to maintain all the components and processes of naturally evolving park ecosystems, including the natural abundance, diversity, and genetic and ecological integrity of the plant and animal species native to those ecosystems."[15]

Wildfire Impacts

National Parks in the western United States are increasingly at risk for high-severity fires. In 2020, 30,000 acres burned in Sequoia and Kings Canyon National Parks, killing more than 7,500 sequoia trees, about 10 percent of the sequoias in the Sierra Nevada. The 2021 Dixie Fire, the largest single wildfire in California history, burned almost a million acres, of which 73,240 acres were in Lassen Volcanic National Park. In 2022 wildfires occurred in Yosemite and Yellowstone, and a wildfire of 18,000 acres in New Mexico threatened Bandelier National Monument and Valles Calder National Preserve. Such events are wholesale change to forest management by reintroducing low-level fire into these landscapes.

---

[15] Page 46, NPS Management Policy, see https://www.nps.gov/subjects/policy/upload/MP_2006.pdf (accessed February 9, 2023).

Native Seed Use

The National Parks use native seed for roadside seeding, revegetation, erosion control, and wildlife habitat. Most seed used in National Parks is collected from the parks by staff and increased by growers under a BPA. The committee could find no source of information on the total scale of use by the National Parks, which operate as independent units. One estimate was that cost of seed collection and increase at a National Park would be approximately $1 million annually, while commercial purchase of native seed will not likely exceed $10,000. SI seed is preferred, and non-native seed is never used. In some cases, Parks have seed-sharing agreements with surrounding counties. In addition, in the past, National Parks have relied on the Natural Resources Conservation Service's Plant Materials Centers (PMCs) for small increases of seed collected from the parks, and for their ability to help growers produce native plants.

Department of Defense

The DOD does not have a direct environmental mission but manages nearly 9 million acres of land in over 440 military installations in the United States (and many more overseas). DOD lands play a significant and perhaps outsized role in maintaining regional and global biodiversity. Many DOD sites contain significant habitats and surviving populations of species in decline or extirpated from other public lands, and over 500 federally listed TES of plants and animals, and at least as many that are considered at risk, more than on any other federal lands. Additionally, the Army Corps of Engineers maintains more than 600 dams, 12,000 miles of commercially navigable inland waterways, and harbors. Much of the Corps' work impacts federal and all other lands, and contributes to the demands on the native seed supply in ways that are significant to this report but are not specifically quantified here.

The DOD has different internal authorities for the use of native seed on military installations, for example, one for remediating disturbances caused by training activities, and another for natural ecological restoration, thus, the size of projects varies greatly. Natural resource managers at the installations operate under the framework of the Sikes Act,[16] passed in 1960, to ensure conservation of fish, wildlife, and natural resources on military lands in the context of ongoing military operations. The Act requires that a collaborative 5-year Integrated Natural Resource Management Plan (INRMP) be developed between installations and the Department of the Interior, typically the USFWS. The INRMP is updated annually and contains a limited number of projects.

Wildfire and Climate Change

The INRMP provides a source of budgetary predictability to the trajectory of restoration activities. However, different sources of funding within DOD are used to address impacts of wildfire, flooding, and other natural disasters each year. Some projects are supported by funds from the USFWS and other agencies. The DOD has developed guidance for its natural resource managers with respect to the incorporation of climate adaptation in the INRMP that incorporates planning, risk assessment, evaluation of implications of options, and adaptive management, including monitoring and adjustment (Stein et al., 2019).

Seed Use

Military installations obtain seed in several ways, including purchases from the commercial market and commissioned collections of seed from installations for both direct use and increase. Because the INRMP is a 5-year plan, advance preparations can be made to acquire seed when needed; however, because some military bases are ecologically unique, some seed types are not available on the commercial market. The DOD has one native seed nursery, on O'ahu, Hawaii, at the Makua Military Reservation.

---

[16] 16 US Code § 670 et seq., as amended.

DISCUSSION

The collective view of the native seed activities of the five large land-management agencies is that of a diversity of approaches to acquiring and using native seed. One key attribute shared by all the federal land agencies is the distributed nature of their decision-making structures. The BLM Field Offices, USFS National Forests, USFWS Refuges, each National Park in the NPS, and individual DOD military bases are the units that ultimately determine seed needs and how to respond to them, what species to use, how much to purchase, and how to implement restoration and revegetation. The federal land agencies also resemble one another in facing several common threats to their natural resource base, in particular the growing frequency and severity of wildfires. Because fires cross jurisdictional boundaries, there is a history of interagency collaboration on projects related to wildfire.

Beyond these similarities, there are also considerable differences among the agencies that are reflected in divergent patterns of procurement and use of native seeds. The USFWS, National Parks, and military bases are focused on activities related to ecological restoration, are generally oriented toward the use of genetically appropriate native seed, and pursue their needs on a project-by-project basis. These agencies appear to have the leeway to seek native seed proactively, or at least to have a longer planning horizon for obtaining the specific native seeds needed for their projects. The INRMP process used by military installations is an example; one of its key features is a strong role for scientific guidance, in that military bases are required to have the oversight of the USFWS on their seed choices. The partnership of the USFWS with local nurseries to provide plants and volunteer organizations to conduct restoration activities also suggest that such projects are implemented through relationship building over time.

In contrast, the two large multiple-use land agencies, USFS and BLM, are concerned with ensuring a supply of seed for a variety of needs. The USFS, with its history of forest commodity production, has developed in-house capacity to meet its own needs for plant materials for ecological restoration (as well as for reforestation) in at least some regions. It is also committed as an agency to ecosystem-based management and has a hard requirement rather than a soft guideline to use genetically appropriate native species in restoration, so the use of non-natives and cultivars in USFS is minimal. Burned forests are allowed to regenerate naturally in many cases.

BLM is committed to a wider range of land uses than USFS, with a greater emphasis on livestock grazing, energy and mineral production, recreation, and other uses, in addition to wildlife conservation and other ecosystem-based activities. Because BLM manages enormous areas of sagebrush, where novel large wildfires made possible by the introduction of cheatgrass are threatening irreversible loss of habitat for sage-grouse and other wildlife, BLM faces a tremendous need to implement post-fire restoration to protect endangered animal species. However, ecological restoration activities within BLM are generally highly intertwined in practice with other activities such as improving forage for livestock. There are no programs within BLM dedicated to ecological restoration as the primary goal, as opposed to being an aspect of programs dedicated to other uses of the land. The expert advisory function is also weakest within BLM; for example, some of its field units and state offices have restoration ecologists, but many do not.

It is very difficult to get a clear picture of the extent of ecological restoration that is pursued by each agency on an acreage basis. The BAR program offers the opportunity to pursue plant community restoration a year or more after a fire, but it is not clear how much restoration has been supported (recognizing that BAR funds have been very limited until they were recently increased through the Bipartisan Infrastructure Law).

Finally, although each of the agencies is working toward building the stocks of native plant seed for their respective needs, their activities do not clearly relate to the methodical fulfillment of the plan put forth to Congress in 2002 to provide for a native plant supply to meet emergency stabilization and longer-term rehabilitation.

FINDINGS FROM SEMI-STRUCTURED INTERVIEWS

This summary of the seed needs of federal land-management agencies is a high-level overview of the current and potential magnitude of native seed uses, and reveals, in part, the directions in which the agencies, or parts of them, are moving with respect to native seed use and capacities. To broaden its understanding of the decision-making process of the agencies, the committee wanted to obtain the bottom-up perspective of agency staff who carry out activities related to native seed use, native seed and plant materials development, research, restoration, and other work.

TABLE 3-3 BLM, NPS, USFWS, USFS, and DOD Uses of Native Seeds (2017–2019)

Purpose of Use
Creation or restoration of wildlife habitat (other than pollinator habitat)
Pollinator habitat projects
Stream erosion mitigation or restoration
Restorative activity on land in a wilderness or natural area
Soil protection
Invasive species suppression
Roadside seeding
Landscaping
Green infrastructure
Roadside maintenance
Natural disaster recovery
Rangeland grazing
Energy development remediation
Green strips (vegetative fuel breaks)

The Committee used a series of semi-structured interviews of personnel from the five major federal agencies that make direct seed purchases. Each respondent talked about the specifics of typical projects involving the use of native seed. This information helped inform the Committee's understanding of the process, and the diversity of paths used in developing and funding a project, selecting and obtaining native seeds for use in that project, and understanding whether the project goals were met. In this section we provide a summary of key topics raised in those interviews. Appendix 2A contains the protocol for the interviews.

These semi-structured interviews of staff from five federal agencies used a somewhat similar line of questions to those asked in the survey of state agency personnel (see Chapter 4). This proved to be difficult to administer because of the widely divergent roles of federal agency staff, which ranged from researchers to natural resource specialists, with many job titles and work locations. The interviews did provide insights about projects within these agencies including the approval process and contracting methods used in purchasing seed, how the seed supplier was identified and selected, and whether the projects had follow-on activities to monitor survival. However, it is important to stress that this process looked at a snapshot of projects with a relatively small number of participants in federal native seed activities and is not considered comprehensive and conclusions should not be drawn to reflect opinions across these agencies.

A list of possible uses of native seeds (Table 3-3) was read to agency staff. They were asked if they had used native seed for any of those purposes in their work.

From the list given to them, no respondent identified green infrastructure as an activity for which native seed was used. But for all other purposes, native seed was used. However, the committee notes that the Army Corps of Engineers, which is increasingly focusing on mitigating the impacts of climate change, stages many projects on and off federal lands that include at least a component of green infrastructure; for many of these efforts, native plant materials are required (Windhoffer et al., 2022). Agency staff noted that leaseholders of grazing permits on federal land were responsible for carrying out agency management prescriptions, which may include seeding activities, with a list of approved plants. It was also pointed out by some respondents that restoration, as a matter of their agency's policy, is conducted on natural areas but not in the wilderness.

When asked for the most common application for using native seeds, the answers included

- Rehabilitation to prevent soil erosion and non-native grass incursion
- Wildlife habitat restoration of invaded habitats, pollinator habitat restoration
- Range improvement and rehabilitation

- Natural disaster recovery including post-wildfire restoration
- Restoration after disturbance (construction projects, military drills, etc.)
- Stream restoration for native salmon
- Reforestation

The typical sizes of projects mentioned ranged from one acre to tens of thousands of acres. The amount of funds spent annually ranged from "no more than $5,000" to more than $1,000,000, which included the cost of seed. Most respondents did not have a good estimate for the amount they spent on seed annually. Topics that arose during various interviews with agency personnel, and some of their observations, can be summarized as follows:

- The populations of native plant communities on land under agency control are important as a supply of seeds for future restoration needs.
- There appears to be a lack of interest by some decision makers in the agencies to pursue ecological restoration relative to other priorities.
- There is a benefit to having a planning horizon of 3 to 5 years, which enables seed to be acquired (or wild collected and increased) and projects implemented according to plans. This allows the ability to anticipate and budget for seed needs annually.
- The typical duration of monitoring the outcome of seeding projects was 2–3 years, though experimental projects (which tended to be smaller in scale) are monitored indefinitely.
- Success in restoration and rehabilitation projects was frequently defined as the amount of "coverage" resulting from the seeding, but this metric may not capture the progressive structure of successful native plant restoration.
- Funding obtained from multiple sources, including grants, funds from the USFWS, and Pollinator Pathways and other sources enable projects that would not happen otherwise. There is no shortage of projects that needed native seed if funding could be found.
- Cooperative efforts exist among botanists, as a group, to prioritize, collect, and develop supplies of seed of non-workhorse species on an annual basis.
- There is a need for more wild collections and increases of native seed to bring more species into use for restoration projects.
- There are labor shortages throughout the agencies, in nurseries, as scouts, and in all manner of fieldwork.
- There is a need for expertise in botany and ecology to inform seed selection and restoration success in different applications. Sources of expertise include the USFWS, PMCs, university extension, and seed producers. Botanists are not equally distributed in all units and regions of the individual agencies.
- There is a need for growers specializing in native plant production for specific environments (e.g., high altitudes or arid) and a need for more growers generally.
- There is a need for more long-term monitoring of restoration efforts to better assess the performance of various natives.
- The performance of natives is not consistent, nor is it *de facto* accepted as superior by everyone.
- There is some ambiguity about the definition of "local" when purchasing seed from the commercial market. Not all seed purchased is certified Source-Identified.

DISCUSSION

The personnel interviewed for this chapter generally expressed satisfaction in the outcome of their seeding projects. Although not necessarily reflective of all personnel in the land-management agencies, many of the interviewees described, as a natural part of their work, how they pieced together different sources of funds to pursue their projects—funds from different authorities within their agencies, funds from other agencies, and funds from external partners, which supported the committee's perception of the lack of dedicated program funding for ecological restoration of native plant communities. The committee observed that native plant communities are the source of seed that underpins many of the activities for which public land is managed, from wildlife habitat creation to oil and gas well pad rehabilitation—and that a central function of land management should elevate the protection and restoration of plant communities.

The committee heard from advocacy groups who commented on BLM challenges in meeting seed needs for restoration. Representatives from Defenders of Wildlife said that there is no single institutional unit that has a budgetary mandate and responsibility for undertaking plant conservation and restoration in BLM, which leads to inadequate long-term planning and accountability for outcomes of seed projects. Advocates from the Western Watersheds Coalition added the need for stronger policy direction to preserve seed sources, perhaps using the Areas of Critical Environmental Concern designation. These speakers described USFS as being more effective in restoration than BLM and attributed this to USFS' adoption of a strong ecosystem management paradigm.

Staff from some but not all agencies pointed out difference in attitudes among decision makers about native seeds, which were attributed to reliance on past practices, or strongly held ideas of what works and what does not, and, with limited funds and time, no inclination to change. All noted that there was much more to be done than funds available. Despite many challenges, agency personnel were optimistic given new federal legislation related to the conservation, restoration, and reforestation of public lands. The personnel interviewed expressed both dedication to land management and concern for the natural resources on public land.

CONCLUSIONS

Conclusion 3-1: The federal government is a major user of native seed, and its uses of native seed have many purposes along the continuum from restoration to rehabilitation and revegetation. Approaches to seed acquisition differ among federal agencies, and these are focused on providing for immediate needs.

Conclusion 3-2: The federal land-management agencies are not prepared to provide the native seed necessary to respond to the increasing frequency and severity of wildfire and impacts of climate change. If this challenge is to be met, the agencies will need to move quickly toward an expanded, proactive effort to develop a supply of native seeds for emergency stabilization and long-term ecological restoration.

Conclusion 3-3: Existing native plant communities on public lands that are managed for multiple uses are an essential resource for developing a native seed supply, but those plant communities are at risk, and their conservation has not been a priority relative to other uses.

Conclusion 3-4: Developing reliable seed supplies for ecological restoration is an achievable goal for large federal agencies, but one that demands substantial institutional commitment to elevating restoration to a high-priority objective, cultivating and empowering ecological expertise to inform their decision making, and creating sustainable funding streams for restoration that support agency personnel to carry out planning, successful implementation, and learning from experience.

REFERENCES

BLM (Bureau of Land Management). 2008. Integrated Vegetation Management Handbook H-1740-2. Available at https://www.blm.gov/sites/blm.gov/files/uploads/Media_Library_BLM_Policy_Handbook_H-1740-2.pdf (accessed January 20, 2022).

BLM. 2020. Public Land Statistics, Volume 205 (June 2021). BLM/OC/ST-21/001+1165. Available at https://www.blm.gov/sites/default/files/docs/2021-08/PublicLandStatistics2020_1.pdf (accessed February 9, 2023).

Bower, A.D., J.B.S. Clair, and V. Erickson. 2014. Generalized provisional seed zones for native plants. Ecological Applications 24(5):913–919.

Breed, M., M. Stead, K. Ottewell, M. Gardner, and A. Lowe. 2013. Which provenance and where? Seed sourcing strategies for revegetation in a changing environment. Conservation Genetics 14:1–10.

CRS (Congressional Research Service). 2022. Wildfire Statistics. IN FOCUS. Available at https://crsreports.congress.gov (accessed February 9, 2023).

Malaval, S., B. Lauga, C. Regnault-Roger, and G. Largier. 2010. Combined definition of seed transfer guidelines for ecological restoration in the French Pyrenees. Applied Vegetation Science 13(1):113–124. http://www.jstor.org/stable/40928281.

Omernik, J.M. 1987. Ecoregions of the conterminous United States. Annals of the Association of American Geographers 77(1):118–125, doi:10.1111/j.1467-8306.1987.tb00149.x.

PCA (Plant Conservation Alliance). 2022. National Seed Strategy Progress Report, Fiscal Year 2021. Washington, DC: U.S. Department of the Interior, Bureau of Land Management.

Smith, F., J. Pierre, M. Young, and D. Devitt. 2020. Estimation of future native grass seed demand for restoring oil and gas-energy sprawl in west Texas, USA. Ecological Restoration 38:237–245.

Stein, B.A., D.M. Lawson, P. Glick, C.M. Wolf, and C. Enquist. 2019. Climate Adaptation for DoD Natural Resource Managers: A Guide to Incorporating Climate Considerations into Integrated Natural Resource Management Plans. Washington, DC: National Wildlife Federation.

US Congressional Research Service. 2020. Federal Land Ownership: Overview and Data. Vincent, C.H. and L.A. Hanson. Report R42346. http://crsreports.congress.gov. Accessed December 1, 2022.

Windhoffer, E., S. Hughes, T. Hughes, A. Grismore, A. Littman, A. Haertling, and J.R. Fischbach. 2022. Consideration of Nature-Based Solutions in USACE Planning Studies. The Water Institute of the Gulf. Funded by the US Army Corps of Engineers Engineer Research and Development Center, Vicksburg, MS.

4

State Government Uses of Native Seed

INTRODUCTION

In the first phase of the assessment leading up to its interim report, the committee heard that native seed shortages in the commercial market were caused by agencies such as the Bureau of Land Management (BLM) buying up all the available seed after wildfires. It put forward the hypothesis that the seed market in the West was strongly affected by decision making of BLM and the US Forest Service, and wanted to know if state governments, which also purchase native seed for land-management activities, were experiencing shortages, and if so, whether this phenomenon was limited to the western states or was more widespread.

The Committee felt that a survey of state departments that use native seed or plant materials might shed light on some of the Committee's other preliminary observations about native seed needs, such as how the timeframe in which seeds are needed and the objectives of users affect availability, among other factors (Box 1-2). The Committee felt that a perspective from state governments would provide another lens with which to assess how well the native seed supply functions across the nation. This chapter presents the survey findings.

Chapter 2 describes the survey methodology. The Committee sent surveys to an average of three government departments in each state likely to use native plant seed and plant materials (to include seedlings, mature plants, and other vegetative materials). As indicated in Chapter 2, this was not a comprehensive list of all state government departments in the United States, so the results cannot be generalized to all such departments. These responses reflect the opinions of the sample that was generated by National Academies of Sciences, Engineering, and Medicine staff based on an internet search of state government webpages to identify those with responsibilities related to land management, fish and wildlife habitat, parks and recreation, roadside vegetation and maintenance, and natural heritage conservation. The departmental staff who responded to the survey were asked to answer questions about the use of seed in the department's work. The staff was also asked about their individual roles related to their department's purchase or use of seed, including project planning, providing biological expertise, purchasing or contracting, project management, or fieldwork (such as site preparation, seeding, and monitoring). Figure 4-1 shows that at least three out of four respondents were involved in planning, advisory, and management roles, while slightly fewer than three out of four were involved in purchasing or fieldwork. The spread across areas shows many respondents appear to have played multiple roles.

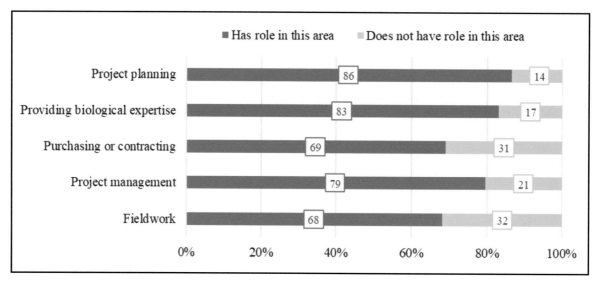

FIGURE 4-1 State survey respondents' role related to the purchase and use of seed and plants, by project area.

Native and Non-native Seed Use

Most state departments surveyed use both native and non-native seeds and plant materials (other than seed) in their projects. Almost 95 percent of respondents to the committee's survey indicated that their department uses native seed or native plants, while three-quarters (74%) of the departments use non-native seed and two-thirds (65%) use non-native plants. About one-fourth of the departments who responded to the survey do not use non-native seed at all and more than a third do not use non-native plants (Figure 4-2).

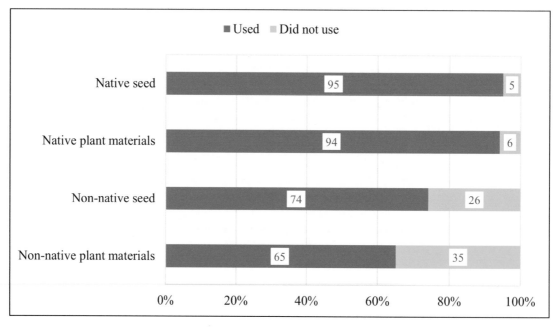

FIGURE 4-2 State departments' use of native and non-native seed and plant materials.

TABLE 4-1 State Departments' Use of Native Seed or Plant Materials for Specific Purposes

Purpose of Use	Percent of Departments
Creation or restoration of wildlife habitat (other than pollinator habitat)	87
Pollinator habitat projects	83
Stream erosion mitigation or restoration	80
Restorative activity on land in a wilderness or natural area	75
Soil protection	72
Invasive species suppression	59
Roadside seeding	67
Landscaping	61
"Green" infrastructure	49
Roadside maintenance	47
Natural disaster recovery	37
Another purpose	31

Most of the survey questions focused on issues related to the use of native seed and plants. To provide a consistent reference timeframe, respondents were asked to focus on projects carried out by between 2017 and 2019.

For example, the survey provided a list of purposes and asked if the department used native seed or plant materials for any of these purposes between 2017 and 2019 (Table 4-1). Appendix 2C shows the exact wording of all of the questions.

When asked about "another purpose" for uses of native seed or plant material, topics mentioned included public education/giveaways, used to restore riparian and wet meadows, provided to producers for seed increase, coastal zone protection, forest management, screening, and the sale of native tree and shrub seedlings for conservation. The survey did not ask if non-native seed or plants were used for these same purposes. When asked if their states have programs to assist private landowners with the use of native seed or plant materials, about half (46%) of the state respondents indicated that they did.

Where Do Departments Obtain Seeds and Plant Materials?

Nearly all the state departmental respondents (96%) who were asked the question said that commercial suppliers are among the sources their department uses to buy native seed or plant materials. Approximately two out of three (63%) said that they obtained native seed or plant materials from project collaborators, and half (49%) said that they collect or grow their own (Figure 4-3).

Collection of Native Seed from State Land and How the Collected Seed Was Used

Figure 4-4 shows that just under half (44%) of the respondents said that the native seed they used between 2017 and 2019 included seed that was wild collected on state land (29% were unsure of whether the seed used had been collected on state land). Of those respondents who responded that wild seed was collected on state land, 92 percent said that at least some of the seed was used to plant directly at a project site, and 79 percent said that at least some of the seed was used to grow plants with the purpose of harvesting seed or plant materials for future use.

When asked about "Other" uses of wild collected seeds from state land, the answers included harvested and given out to landowners, grown into seedlings for sale to residents, frozen for storage or given to seed banks for future use, and used for projects on partner lands.

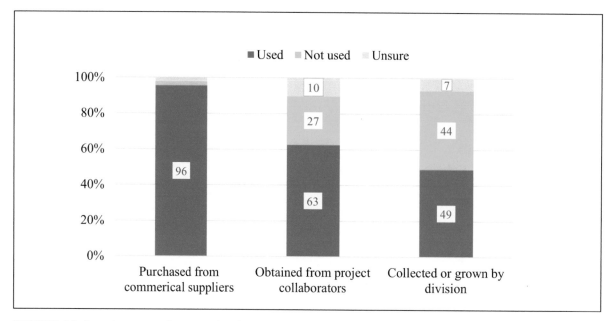

FIGURE 4-3 State departments' sources of native seed and plant materials.

Ambient, Refrigerated, and Frozen Storage

Respondents were asked about the availability to their departments of different types of seed storage. Sixty-four percent said they had ambient storage space, 28 percent had refrigerated storage, and 16 percent had freezer storage of some kind (13% were unsure about whether they had refrigerated storage and 15% were unsure about freezer storage).

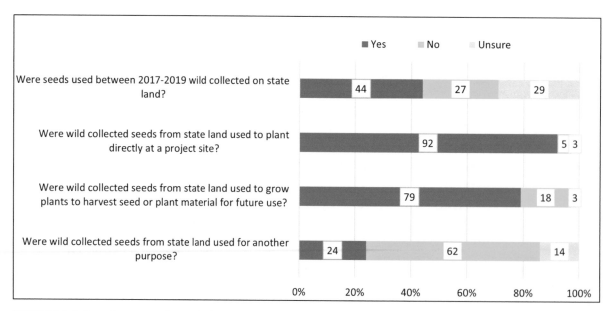

FIGURE 4-4 State departments' use of wild-collected seed from state land.

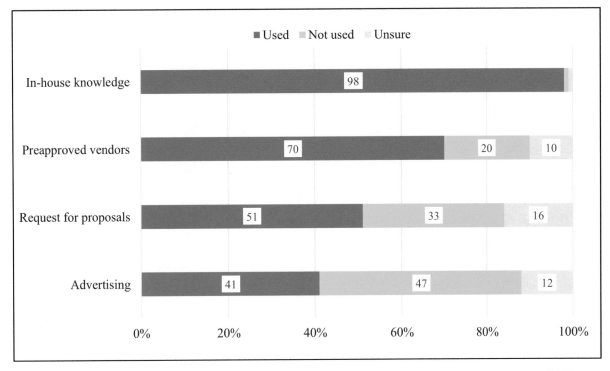

FIGURE 4-5 Sources of information used by state departments to learn about native seed and plant material availability.

Sources of Information about Native Seed and Plant Materials Availability

Respondents were given a series of questions asking whether a given option is used as a source of information to find out about the availability of native seed and plant materials. The options included in-house knowledge, advertising (about needs), preapproved vendors, or requests for proposals. Figure 4-5 shows that 98 percent of departments rely on in-house knowledge to find the seed and plant materials. Preapproved vendors and Requests for Proposals are used by most departments, while advertising is the least common method.

Departmental representatives were asked if there were other sources of information used to obtain information about native seed and plant materials availability. Responses included direct calls to vendors, annual meetings with registered suppliers, online vendor postings, other state departments, state crop associations, state university extension offices, native plant databases, BLM seed buys, and state seed bank.

Relationships with Suppliers

How Seeds Are Obtained from Commercial Suppliers

Among those who said that their departments bought native seed or plant materials from commercial suppliers, 81 percent of the respondents said that this involved purchases directly on the open market, and 66 percent said that purchases were made via a formal bidding process. To gain a sense of the differences between eastern and western states, the responses were compared by region. As described in Chapter 2, the survey included a relatively small number of departments, and therefore the comparisons by region should be interpreted with caution.[1] Figure 4-6 shows

---

[1] East includes AL, AR, CT, DE, FL, GA, IA, IL, IN, KS, KY, LA, MA, MD, ME, MI, MN, MO, MS, NC, ND, NE, NH, NJ, NY, OH, OK, PA, RI, SC, SD, TN, TX, VA, VT, WI, WV; West includes AK, AZ, CA, CO, HI, ID, MT, NM, NV, OR, UT, WA, WY.

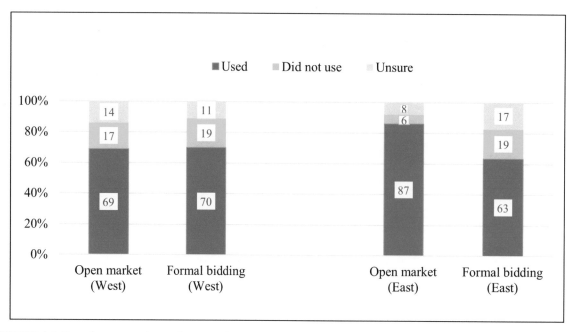

FIGURE 4-6 State departments' use of open market and formal bidding when purchasing native seed or plant materials from commercial suppliers, by region.

that in the eastern states, a larger percentage of departments appear to use purchases from the open market (87%) as compared to 69 percent in the western states. In addition, 63 percent of respondents located in the eastern states said that they use the bidding processes while this value was 70 percent in the western states.

Marketing Contract and Production Contracts

Respondents were asked about contracting arrangements, and specifically whether their department used marketing and/or production contracts to procure seeds from suppliers. Marketing contracts specify the type, price, quantity, and delivery date of seed or plant materials. By contrast, production contracts specify the desired type, quantity, and delivery date of native seed and plant materials, while mitigating production risk by sharing some production costs or by providing flexibility on the quantity delivered and/or the delivery date. They may or may not specify a price. Only 22 percent of departments said that they used production contracts, compared with 39 percent who used marketing contracts (Figure 4-7). Marketing contracts may offer more certainty to buyers and sellers, but production contracts might be used to encourage growers to take on the production of a difficult-to-grow species. About 18 percent of state respondents were unsure if marketing contracts were used and about 25 percent were unsure if production contracts were used. This is consistent with the variety of different positions and responsibilities held by respondents who may lack individual knowledge about the purchasing process.

Timing of Contracts Relative to Production

Two out of five respondents from the state departments were unsure about whether the contracts are established before or after the supplier begins the seed production process. Thirty-seven percent said that the contract was usually established before the supplier begins the seed production process, and 24 percent said that it was after (Table 4-2).

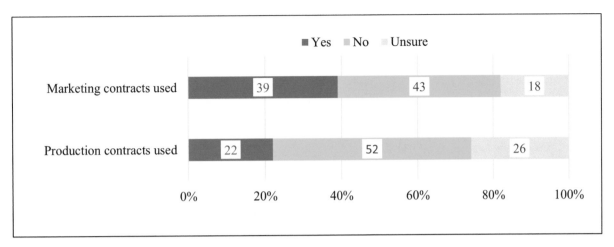

FIGURE 4-7 State departments' use of marketing and production contracts.

Communicating Future Seed Needs to Suppliers

No single method dominated the state departments' methods of communicating their anticipated future needs to suppliers. Approximately half of the respondents mentioned word of mouth (55%), requests for proposals (49%), conferences or other professional meetings (49%), and informal meetings with growers (45%).

Importance of Seed Attributes

Departmental representatives were asked to reflect on projects between 2017 and 2019 and to indicate how important it was to obtain seed with the following attributes: native seed; native seed sourced from a specific geographical location or seed zone; and certified native seed. Nearly all departments nationwide indicated that native seed was very important or somewhat important. As Figure 4-8 shows, a higher proportion of departments in western states indicated that native seed was very important (90%) relative to departments in eastern states (79%). Departments in western states were also more likely to view native seed sourced from a specific geographical location or seed zone as very important (57%) relative to eastern states (46%). The responses with respect to certified seed were similar across the country: 42 percent of departments from East and 43 percent of departments from the West indicated that certified native seed was very important.

Responses were also compared based on the departments' average annual expenditures on native seed and plant materials. Similar to the geographic comparisons, the comparisons by the level of expenditure should be interpreted with caution due to the small sample size. In addition, approximately 10 percent of the state respondents were unable to provide an estimate of the amount their department spends on native seed and plant materials, and some respondents skipped this question.

TABLE 4-2 Timing of When State Departments Usually Establish a Contract with Seed Suppliers

Contract is Usually Established	Percent of Departments
Before the supplier begins seed production	37
After the supplier begins seed production	24
Unsure of timing of contract	39
Total	100

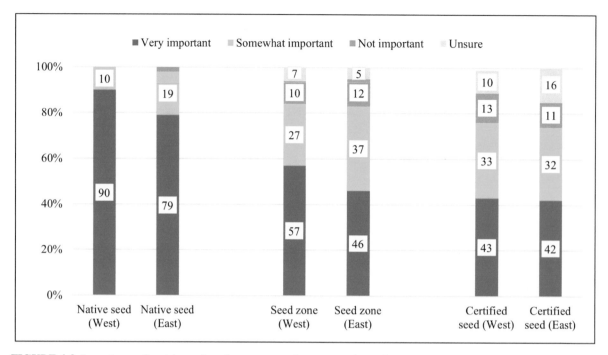

FIGURE 4-8 Importance of certain seed attributes to state departments, by region.

When responses were examined relative to the level of departments' annual expenditures on native seed and plant materials, those with larger annual expenditures (over $100,000) were generally somewhat more likely than those with small annual expenditures ($100,000 and under) to indicate that each of the three seed attributes were very important. Figure 4-9 shows that certified seed was very important to departments with larger expenditures (54%) relative to departments with smaller expenditures (31%).

In general, the results show that using native seed was more important to states than using native seed sourced from a specific geographic location or certified seed (in which the geographic source of the originally collected seed was verified

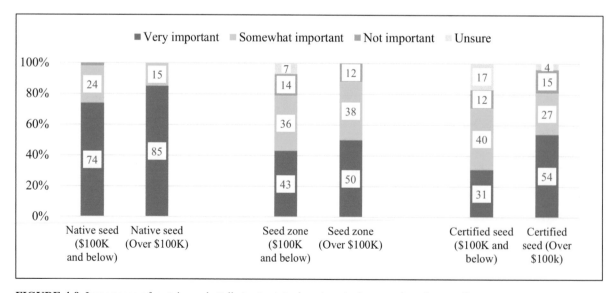

FIGURE 4-9 Importance of certain seed attributes to state departments, by annual seed expenditures.

TABLE 4-3 Other Seed Attributes Specified as Important by State Departments

Desired Attributes	Number of Responses
Genetic appropriateness	7
Weed-free	6
% Purity	6
High germination rate	6
Prior track record	6
Other	6
Cost	5
Total	42

by a certification agency, along with the genetic integrity of subsequent generations of increase. Seed certification is not required for the sale of seed (in contrast to seed viability analysis; see Chapter 8), but it is sometimes required by buyers. The cost of certified seed is a little more than non-certified seed because of the fee associated with certification.

Departments were asked to list other characteristics of seed obtained during 2017–2019 that were important. Table 4-3 is a summary of responses received.

Attributes mentioned related to basic market conditions included cost (5), quantity (2), and availability (2). Cost corresponds to price (cost per unit), or price multiplied by quantity (total cost). Many other attributes mentioned regard quality dimensions that are independent (or potentially independent) of source specificity, e.g., weed-free, with genetically appropriate mentioned most often.

Seed Substitutions

State representatives were asked questions about how often their departments substituted other kinds of seed or plant materials when their preferred native seed or plant material was unavailable, specifically substituted with non-native seed or plant material, native seed or plant material of another species, or native seed or plant material from a different geographic region. In some cases, the responses to different kinds of substitutions varied regionally or according to the level of a department's annual seed expenditures. Figure 4-10 shows how departmental representatives across all states answered these questions.

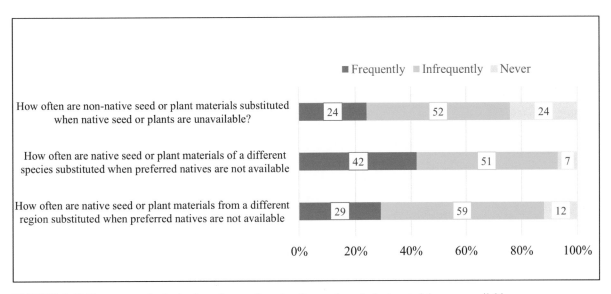

FIGURE 4-10 Frequency of substitution when preferred native seeds and plant materials are unavailable.

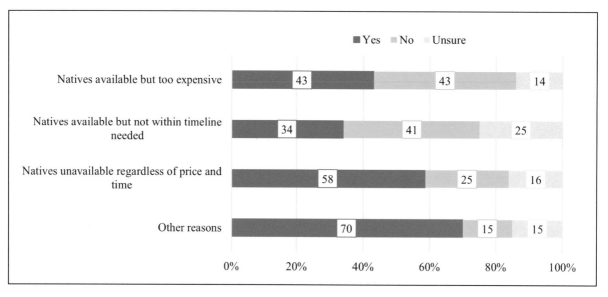

FIGURE 4-11 Typical reasons for substituting with non-natives.

Substituting Non-Natives for Natives

The responses of state representatives, asked how often their departments substitute non-native seed or plant materials when native seed or plants are unavailable, was examined by region and by the level of annual expenditures of a state department ($100,000 and below, or over $100,000). The responses were similar by region and size and the overall national numbers. About a quarter (24%) of departments frequently substitute with non-natives, and about half said that they substitute non-natives infrequently. A quarter (24%) indicated that their agency never substitutes non-natives for natives, even if natives are unavailable.

Respondents from those departments that said that they substituted non-native seed or plant materials (either frequently or infrequently) were asked about several potential reasons for the substitutions: native seed or plant material were available but too expensive; native seed or plant material were available but not within the timeline needed; and native seed or plant material were unavailable regardless of price and timing. Figure 4-11 shows the agency responses overall. The most commonly cited reason by state departments overall (58%) was that native seed or plant material were unavailable regardless of price and timing. The second most cited reason was that preferred natives were available but too expensive (43%). Timing considerations were cited by 34 percent. Approximately two out of five respondents cited at least two of these three market-related reasons, and a fifth of respondents cited all three reasons, meaning that they cited price and timing reasons as well as "unavailable regardless of price and timing." One potential explanation for this finding is that some native seed species were entirely unavailable, while others were available but costly or at the wrong time.

Figure 4-12 shows responses by region to the questions about reasons for substitution with non-natives. Departments in western states found cost, availability, or both to force a substituting with non-natives 50 to 70 percent of the time, versus 26 to 52 percent in the East. Notably, 70 percent of departments in the West encountered situations when native seed or plant material were unavailable regardless of price and or in the timeframe needed.

Other Reasons for Substituting with Non-natives

In an open-ended question, state staff were also asked if there were other reasons besides the three offered by the survey that explain why non-native seeds and plant materials were substituted when preferred natives are not available. Table 4-4 provides a summary of the reasons given, grouped into categories.

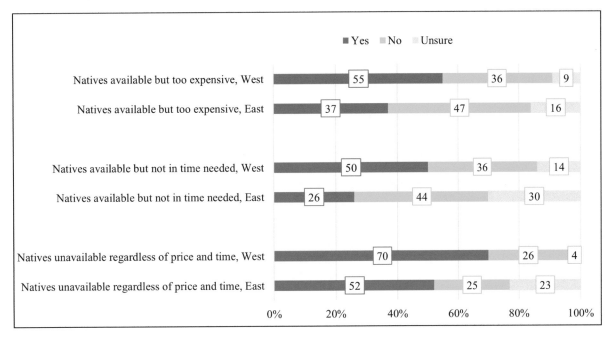

FIGURE 4-12 Typical reasons for substituting with non-natives, by region.

Substituting with Natives of a Different Species or from a Different Region

When preferred native seed or plants are not available, departments sometimes substitute with native seed or plant materials having different characteristics. As shown in Figure 4-13, 42 percent of the respondents stated that they frequently substitute with native seed or plant material of different species, and this was similar for eastern and western states. When the responses were examined by the level of annual expenditure on seeding projects, 50 percent of state departments with $100,000 or more in expenditures indicated that they frequently substitute with

TABLE 4-4 Other Reasons Provided by State Departments for Substituting Non-Natives for Preferred Natives

Reason for Substitution (Categories)	Description of the Types of Reasons Given in Each Category	Number of Responses
Effectiveness	Non-native seed or plant materials are associated with better performance with respect to areas prone to erosion, such as near infrastructure or buildings, on steep slopes, for soil stabilization. Non-natives are associated with a stronger ability to compete with invasive species. Non-natives meet requirements to provide a % ground cover. Better performance of mixtures.	10
Specific purpose	Non-native seeds and plants are specifically selected for ornamental and fruit trees, landscaping, and agricultural plantings, and for wildlife food/forage.	7
Preference for non-natives	Land manager discretion, preference, previous experience, lack of knowledge of natives, lack of options	6
Other	Seed offered was not certified, insufficient quantities available, poor seed quality/viability, lack of ecoregional genotypes, no inexpensive native forbs	5
Total		28

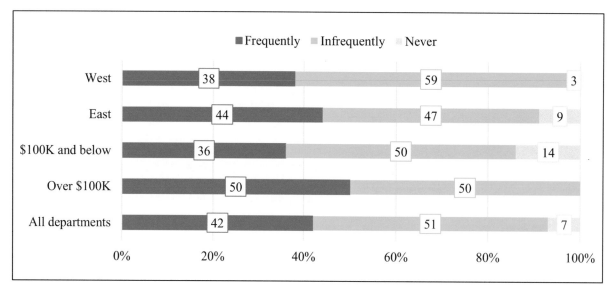

FIGURE 4-13 Substitution with natives of different species when preferred natives are unavailable, by region and annual seed expenditures.

natives of different species versus 36 percent of departments with expenditures of $100,000 or below. This suggests that those departments with larger projects may have more difficulty getting the quantity of natives they need.

As shown in Figure 4-14, 29 percent of state representatives responding to the survey indicated that their departments frequently substitute with native seed or plant materials from a different geographical source location when their preferred native seed or plant materials are not available. Examined on a regional basis, 45 percent of respondents from western states substitute with natives from a different geographical location, versus 21 percent from eastern states. One possible explanation for this difference is that there is a greater diversity of landscapes and seed zones in the West so it may be harder for suppliers to provide a sufficient diversity of ecotypes to meet all needs. The responses were similar across state departments with different levels of annual expenditures on seed.

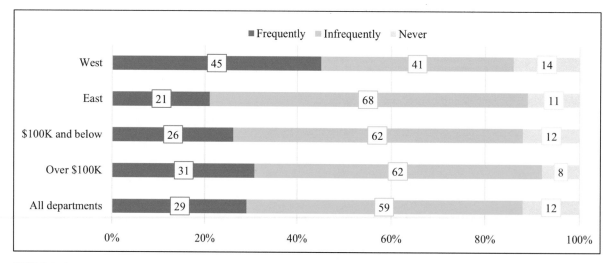

FIGURE 4-14 Substitution with natives from different region when preferred natives are unavailable, by region and annual expenditures.

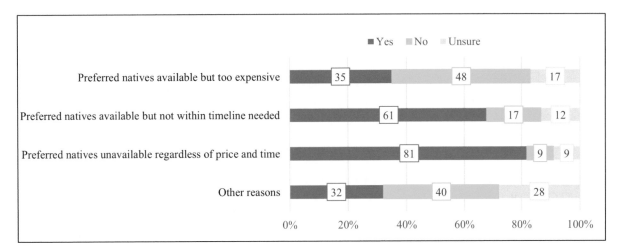

FIGURE 4-15 Typical reasons for substituting with natives having different characteristics.

Among the reasons for substitutions with natives having different characteristics, the lack of availability regardless of price or timing was mentioned by 81 percent of respondents overall, lack of availability within the timeframe required by 61 percent, and price by 35 percent (Figure 4-15). All three reasons were cited by 25 percent of respondents. Unavailability was cited in conjunction with one of the other reasons more than half of the time. Thus, the respondents citing unavailability and at least one other reason are a substantial share of the 81 percent reporting unavailability as a reason. Almost half of respondents (48%) indicated that the price of available seed was not a reason.

Figure 4-16 shows that the lack of availability appears to be a bigger challenge for departments in the West than the East. Lack of availability in the time needed was mentioned by a higher proportion of respondents in the West (79%) versus the East (51%). It is possible that buyers in western states have smaller windows of time for purchasing seed

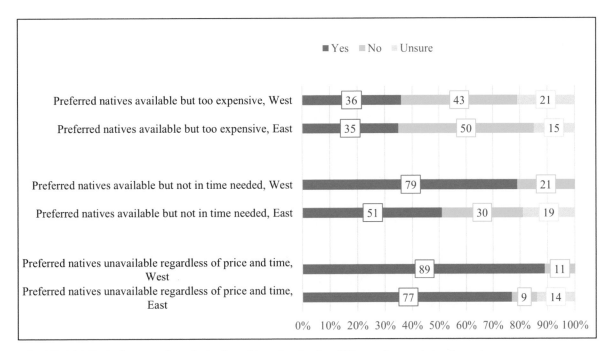

FIGURE 4-16 Typical reasons for substituting with natives having different characteristics, by region.

than those in the eastern states, due to the need to obtain seed for emergency stabilization after wildfire, for example. About one in three staff in both the East and West indicated that substitutions with native seeds of other species or from other regions were made because the seeds they wanted were available but too expensive. Eighty-nine percent of state staff in the West and 77 percent in the East said that substitutions were made because the preferred seeds were unavailable regardless of price and timing. Departments with seed expenditures over $100,000 were more likely (92%) to say that substitutions with native seed with other characteristics were made because preferred seed were unavailable regardless of price and timing than departments with seed expenditures of $100,000 and under (78%) (data not shown).

Biggest Barriers to Using Native Seed

When asked (in the form of an open-ended question) about the biggest barriers or disincentives to using native seed and plant materials in the state departments' work, approximately half of the respondents mentioned availability of plant material (48%). The second most frequently mentioned barrier or disincentive was price (38%). These two reasons are interdependent in a market. If buyers, such as these survey respondents, are unwilling or unable to pay a price that at least covers the suppliers' cost of production or collection for some species then none of the desired seed will be produced. In turn, buyers will experience a lack of availability. Because nothing is produced, buyers will perceive the outcome as "unavailable regardless of price and timing." Likewise, if buyers fail to signal their intent to purchase seed of designated species in time to produce that seed, suppliers may deem it too risky to incur production costs with the market in doubt. Both of these reasons for unavailability result from market forces.

Expertise and Labor

Monitoring Seeding Projects

Most state respondents said that they check on the survival of the seed or plant materials after planting for most or some of their projects (72% do this for most projects and 26% do this for some projects). These responses were also examined by region: 80 percent of the respondents in the West said that they do this for most projects, compared to 67 percent in the East. As shown in Figure 4-17, 27 percent of departments in the East check on survival of seeding or

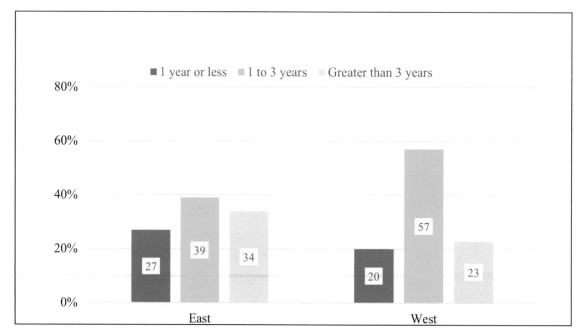

FIGURE 4-17 Time of last check on the survival of seed or plant materials after planting for a typical project, by region.

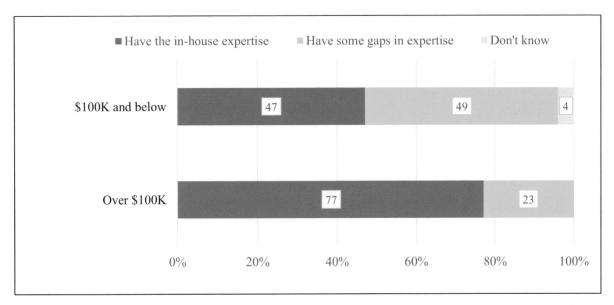

FIGURE 4-18 Availability of in-house expertise for projects that use native seed and plant materials, by annual expenditures.

planting projects for a year or less, versus 20 percent of departments in the West. A final survival check was typically completed within 1–3 years by 57 percent of departments in the West versus 39 percent in the East. About a third of departments in the East and about a quarter of departments in the West continue to monitor seedings and plantings for survival for greater than 3 years.

In-House Expertise and Gaps in Expertise

Department representatives were asked whether they have the range of in-house expertise needed for the various aspects of projects that use native seeds or plants. Overall, 58 percent said that they have the expertise needed, and 39 percent said that they had some gaps in expertise (the remaining 3% did not know). Responses to this question did not differ by region (57% said that they had the needed expertise in the West and 59% said the same in East). Responses were also examined by the amount of annual native seed and plant materials expenditures of the department (Figure 4-18). Those with larger annual expenditures were more likely to have in-house expertise than those with smaller expenditures (77% of the departments with budgets over $100,000 said that they had the expertise compared to 47% of the departments with budgets of $100,000 and under).

Those who said that they had some gaps in in-house expertise were asked to describe the expertise they were missing. In addition to mentions of inadequate levels of staffing, the responses covered a broad range of gaps in specialized knowledge, including agronomy, seed production, propagation protocols, and collection methods, native plant species identification, and planning for seeding with natives and maintenance. Some respondents said that they were able to meet their needs with various forms of collaborations with outside experts, and some clarified that the gaps only affect certain parts of the state.

Role of Policy in Decision Making

Most respondents said that a variety of specifications and guidelines apply to their projects that involve native seed and plant materials, including technical specifications (91%), federal regulations, guidelines, and policy (71%), state regulations, guidelines, and policy (84%), and funding source specifications (57%). The survey did not ask the extent to which these factors influence decision making, and respondents did not bring them up in the open comment section.

Barriers to Using Native Seed

In answers to an open-ended question, state representatives provided perspectives on the key barriers that prevent them from using native seeds in their department's work. Table 4-5 summarizes the count of responses, coded into categories.

Availability and cost of genetically appropriate native seed were the concerns most strongly expressed by respondents to our survey of state seed buyers. Project timing was also cited as a limitation for obtaining sufficient supply. Desired plant materials are often unavailable in the quantities needed, and buyers' project timelines do not allow for the time that it might take suppliers to obtain and/or propagate these materials. High-quality native seed may also be prohibitively expensive, and even when state departments have a policy of using it, their project contractors may substitute lower-cost seed. The interrelationship of price and availability through the native seed market indicates that basic market considerations drove most responses. However, especially in the West, native seed with the species and traits needed was sometimes unavailable at any price or time.

Lack of support in the context of the state bureaucracy was also cited by some respondents. Ecological restoration and the use of locally adapted native seed are not seen as a high priority in some state departments, leading to a lack of clout to improve the process or increase the funding for seed purchases. State agency staff or their project contractors may lack the technical knowledge to carry out restoration with native plant species. However, some state respondents reported the perception that the mandate to use native species in restoration is increasing in their agency.

Some of the respondents commented that native species do not tend to meet their agency's regulatory mandates for successful post-disturbance revegetation. Desirable attributes that native species may lack include establishing rapidly, meeting a certain percent cover in a specified time, preventing erosion on steep slopes, persisting under automobile pollution, tolerating degraded soils, inhibiting weeds. and resisting pests and diseases. These responses

TABLE 4-5 State Department Respondent Comments on Barriers to Using Native Seeds

Barriers (Category)	Description of the Types of Comments Included in Each Category	Number of Responses
Availability	The lack of genetically appropriate or local ecotypes of natives in large quantities in the timeframe needed is a major barrier. The period in which funding is available does not match the time needed to contract out the production of appropriate seed.	48
Cost	The costs of natives are likely to be prohibitive in many cases unless natives are mandated by policy. Project contractors often substitute lower-cost seed or mixes.	38
Genetics	A source of genetically diverse, locally adapted ecotypes of native plants is not available on the open market.	16
Lack of support for native plants/seeds	Lack of support for native plants and/or an established reliance on non-native species. Lack of support could be from engineers, landowners, public, etc. The historical reliance on non-native plants makes the transition to native plants difficult. Respondents mention a kind of "mentality" or suggestions of a certain mindset.	13
Lack of knowledge	Staff do not know how to establish or maintain native plant communities. On roadsides inappropriate mowing and use of herbicide are frequent.	11
Geographic or ecosystem limitations/ conditions	Natives often do not perform well at meeting regulatory specifications: establishing rapidly, meeting a certain percent cover in a specified time, preventing erosion on steep slopes, persisting under automobile pollution, tolerating degraded or compacted soils, inhibiting invasion, and resisting pests/ diseases. This is an important research need.	9
Timing	Project timelines do not allow for the delays inherent in obtaining genetically appropriate native plant material, nor for the time it takes to monitor outcomes.	9
Procurement barriers	State procurement systems are cumbersome when coupled with a lack of understanding about native seeds.	5
Communication issues	There is not enough native seed expertise throughout the agency to be able to communicate what kind of seeds are to be used in seeding projects. The definition of what is "native" is more restrictive in some parts of the agency than others.	1

may be a reflection of the fact that the state survey included departments of transportation and departments of natural resources, parks, and wildlife, which are somewhat divergent in their interests and perspectives.

Prospects for Native Seed and Plant Material Use

The majority of respondents anticipate that their department's use of native seed or plant materials is likely to increase in the future, both near term (76%) and long term (73%). In answers to an open-ended question, state agency representatives explained why they thought the use of native seed or plant materials by their departments is likely to increase or decrease. A high-level summary of responses to perspectives in the short term is provided in Table 4-6 and in the long term in Table 4-7.

TABLE 4-6 Reasons for Anticipated Increase or Decrease in State Departments' Near-Term Use of Native Seed or Plant Materials

Reasons Native Seed and Plant Materials Use Is Likely to INCREASE in the Near Term (Category)	Description of the Types of Comments Included in Each Category	Number of Responses
Preference, demand, interest	There is increasing preference for native seed/plants over non-native seed/plants. Demand and interest for native seed/plants is growing.	13
Habitat programs, restoration	The number of habitat and/or restoration projects that will use native seed/plants is growing both in the public and private sectors.	12
Awareness, education	There is increasing awareness of the public and more public education about the benefits of using native seeds/plants.	11
Environmental reasons	The ecological and environmental benefits of using native seed/plants are becoming clearer.	8
Projects	Construction projects, road-building, and other activities will necessitate seeding projects.	7
Research, funding	Research and funding are beginning to increase to allow for further investment in native seed/plants.	6
Relationships	Relationships with suppliers to collaborate on growing native plants and partnerships that support investment in, and further development of, native plant/seed programs, such as pollinator programs.	4

Reasons Native Seed and Plant Materials Use Is Likely to DECREASE in the Near Term (Category)	Description of the Types of Comments Included in Each Category	Number of Responses
Staffing	There are not enough staff to undertake the work at every level, from writing grants to obtaining funds for native seed projects, to actually managing those projects.	1

Reasons Respondent is UNSURE About Whether Native Seed and Plant Materials Use Is Likely to Increase or Decrease in the Near Term (Category)	Description of the Types of Comments Included in Each Category	Number of Responses
Natives already	The existing seedbank of natives is the main source of natives that is used.	4
Funding dependent	Many seeding projects are cost-shared with private landowners, and some in conjunction with USDA conservation projects.	3
Remain the same	Given current project needs, the use of natives will be similar in the future.	3
Climate dependent	Persistent drought will make the use of native seed more difficult.	1
Innovation dependent	Some environments, soils, slopes are not conducive to the establishment of native vegetation, but new seeding techniques may help.	1

TABLE 4-7 Reasons for Anticipated Increase or Decrease[a] in State Departments' Long-Term Use of Native Seed or Plant Materials

Reasons Native Seed and Plant Materials Use Is Likely to INCREASE in the Long Term (Category)	Description of the Types of Comments Included in Each Category	Number of Responses
Preference, demand, interest	Legislative mandates, habitat programs, conservation efforts, and new initiatives to create partnerships to develop native seed supplies will support long-term use.	13
Habitat programs, restoration	The number of habitat and/or restoration projects that will use native seed/plants is growing both in the public and private sectors.	9
Environmental reasons	Resilient ecosystems will adapt more readily to climate change and native plants contribute to resilient soils, wildlife habitat, and ground cooling.	7
Awareness, education	Recognition that native plants will be lower maintenance in the long run will get more people on board.	7
Same reasons as for short-term increase	Same reasons as for short-term increase.	5
Research, funding	There will continue to be an increase of funds for restoration due to Monarch butterflies, pollinators, endangered species, and efforts to build a stock of native plant seeds/materials.	4
Population growth	Increase in numbers of people moving into former agricultural areas will likely mean shifts in land use that will favor conversion and use of natives.	2

Reasons Respondent is UNSURE About Whether Native Seed and Plant Materials Use Is Likely to Increase or Decrease in the Long Term (Category)	Description of the Types of Comments Included in Each Category	Number of Responses
Markets	Economic markets may determine how private land is used.	3
Climate dependent	Global warming and natural disasters will play a role in ability to successfully restore native plant communities.	2
Needs dependent	Establishment of plant communities is difficult, but in the long term a move from establishment to maintenance could be possible.	2
Funding, research dependent	There are so many needs for restoration and so many research needs to help make those successful. If research funding is available to address these issues, the use of natives will continue.	1

[a] None of the respondents provided a reason why native seed and plant materials would likely decrease in the long term.

DISCUSSION

Different departments within one state have restoration projects that can be very diverse in purpose, making it difficult to represent the view of "the state" with any one answer, so the responses to the survey may be a partial picture of state native seed users. Some of the respondents noted that different answers might come from a more targeted approach to different departments within their state. Thus, this survey should be considered a first step in elucidating the seed needs of states.

The survey showed that there is growing recognition of native seed and plant materials by states as being valuable for their environmental benefits, from improving water quality to providing wildlife habitat, and the use of native seeds by states is expanding. Although the survey found that virtually all state departments say it is important to use natives, not everyone in state departments agrees that native plants are the best choice for their restoration projects. They can be difficult to grow, there is a deficit in the knowledge base about how to grow native plant communities, and there is historical reliance on non-natives.

States are on a learning and experience curve in forging a common understanding about the use of natives. Respondents made the point that proponents of native seeds should recognize that native seeds and plants have limitations in very altered environments, in which pollution, salt operations (on roads), compacted soils, and invasive plants, pathogens, and pests make native plant communities unsustainable. They cautioned that the debate about "what is native in a region" tends to be so granular as to work against encouraging the broader use of natives including cultivars.

The survey results showed that most states have policies that require the use of natives, but respondents suggested that factors such as "greater awareness," "scientific evidence," and "availability" were more likely than policy (which often has loopholes) to play a role in increasing the use of native seeds and plant materials by department staff. The survey found that all states prefer natives but having local ecotypes of natives is a little less important and having the source location of seed verified by certification is slightly less important. On the other hand, non-certified seed obscures the geographic origins of plant materials and may lead to the use of inappropriate seed and materials for a region, leading to restoration failures that reinforce a perception that natives do not work. Many respondents stated that seed and seed mix selection guidance would be helpful.

The survey results suggested some differences between states in the East versus West, keeping in mind that the results need to be interpreted cautiously, given the small numbers of participants in the survey. Respondents from western states had more difficulty than their counterparts in the East finding their preferred seed types regardless of timeframe needed or price, which might be related to the larger number of seed zones in the West than in the East, and the fact that obtaining seed from a particular seed zone was more likely to be very important to state respondents in the West than the East. More western state respondents said that their preferred seeds were also not available in the timeframe in which seed was needed, which might reflect a shortage during the years of frequent wildfire, when seeds are being sought in large quantities by BLM and other agencies.

States in the East were more likely to buy off the shelf from seed suppliers than request a formal bid. All states were less likely to engage in production contracts than marketing contracts, but many respondents in the survey were not familiar with the procurement details of seed acquisition. In open comments, some respondents said there is difficulty in using project funds intended for immediate needs to try to collect and store seeds for increase for future needs. The mismatch between time constraints on the use of funding and seed production was noted as a major barrier to the native seed supply.

Finally, some states have a good supply of existing native plant communities which, if maintained, can be the basis of a seed supply for restoration, and there is hope to be able to use those native stands effectively, without needing to buy seeds.

CONCLUSIONS

Conclusion 4-1: *Virtually all state departments prefer to use native seeds in their seeding projects relative to non-native seeds, with a few exceptions, and most states have policy guidance requiring natives to be used. Obtaining native seed from a specific geographic source or seed zone was also important to a majority of departments, but less so relative to the importance of using native seeds. A slightly smaller percentage of department respondents said that obtaining certified seed was very important for their department, relative to using native seed or seed from a specific geographic origin or seed zone.*

Conclusion 4-2: *Many state departments appear to have faced difficulty maintaining a diverse supply of genetically appropriate native plant seeds and plant materials and consequently substitute non-natives, different species, or the same species of plant but from another region. The major reasons cited for substitutions are a lack of availability at any price, and the price, when seed is available.*

Conclusion 4-3: *Almost all the state departments obtain seeds from commercial suppliers and about half of the departments collect seed from state lands or grow their own, primarily for use on state lands. The timeframe during which funds are available and must be spent is mismatched to timing of seed production. This was noted* *as a major barrier to the native seed supply. Ambient storage for seed is not available in all states, and less than a third have refrigerated storage available.*

Conclusion 4-4: *The lack of knowledge on establishing native plants was mentioned as a barrier to native seed use. Seed selection tools, greater knowledge about restoration outcomes, and better information about the performance of native seed mixes used in different applications would be helpful to states.*

Conclusion 4-5: *There is growing recognition of native seed and plants as valuable for their environmental benefits, from water quality to wildlife habitat, but procurement processes and lack of cooperation by some departments and private-sector contractors undermine the use of high-quality natives.*

5

Tribal Uses of Native Seed

HISTORICAL FACTORS AFFECTING TRIBAL NATIVE SEED CAPACITIES

Tribal needs concerning native plants often parallel those of federal and state agencies, and private enterprise described in this report including seed collecting, increase, testing, seed zone implementation, storage, contracting, and dedicated funding. Still, tribes face unique issues that complicate actions needed to enhance native plant needs and uses. These include historical federal land policies and complex government-to-government interactions. In addition to ecological restoration, native plant uses also include food, spiritual, and medicinal applications (Mike et al., 2018).

The Dawes Act, also known as the Allotment Act of 1887, resulted in the loss of two-thirds of reservation land deemed "in excess of Indian needs."[1] It also assigned small land allotments within reservations to tribal members. The allotted land could be sold to non-tribal entities and so privately owned land became scattered within reservation borders and is commonly referred to as checker boarding. Additionally, there has been a fractionalization of land by the legal requirement that inheritance be equal and undivided among heirs and over generations, resulting in some parcels with hundreds of owners (Murray, 2021).

After the federal cultural assimilation policies ended with 1975 Indian Self-Determination and Education Assistance Act, tribes have sought and obtained new levels of self-determination and sovereignty, and pursued active reclamation and maintenance of traditional culture, including native plant uses (Mike et al., 2018). Still, land management is a balancing act between tribal sovereignty, complicated land ownership categories, and the Bureau of Indian Affairs' (BIA) legal and fiduciary obligation under the Federal-Tribal Trust to protect tribal treaty rights, lands, assets, and resources.[2] Limited resources and the multifaceted interactions of tribal governments with federal and state governments, and the BIA, continue to complicate land-management actions in general, including native plant uses for restoration and cultural needs.

Although viewpoints on land use, conservation, and management differ widely among and within tribes, there are unifying aspects of conservation and use of native plants. With enhanced self-determination, tribal nurseries supporting cultural uses and ecological restoration have been strengthened over the last 20 years by the Inter-Tribal Nursery Council (INC). There are also efforts to expand tribal conservation districts in partnership with the Natural

---

[1] See https://www.archives.gov/milestone-documents/dawes-act (accessed October 23, 2022).
[2] See https://www.bia.gov/bia (accessed October 23, 2022).

Resources Conservation Service (NRCS).[3] These districts can assist tribes and tribal members in agricultural production, including growing native seed for tribal use and general sales. Since indigenous people traditionally lived close to the land and its resources, they developed management regimes over centuries to secure and benefit their livelihoods (Reyes-Garcia et al., 2019). This has led to the recognition and implementation of tradition ecological knowledge as an asset to tribal land management and ecological restoration (Eisenberg et al., 2019).

THE INTER-TRIBAL NURSERY COUNCIL

Jeremy Pinto, a US Forest Service (USFS) Research Plant Physiologist, presented information about the Inter-Tribal Nursery Council to the committee. Starting in 2001, tribal emphasis and outreach programs were developed within the Reforestation, Nurseries, and Genetics Resources (RNGR) program,[4] a USFS-sponsored effort to supply growers with needed technical information for growing native plants. The first meeting of tribal nurseries under the aegis of RNGR was in 2001 at Fort Lewis College in Durango, Colorado, which led to the establishment of the INC. The Council is critical to improving nursery operations and management, and to strengthening connections among tribal nursery leaders. The INC meets annually to focus on topics that include technology transfer, conservation, education, traditional ecological knowledge, and general restoration.

The Tribal Nursery Needs Assessment (Luna et al., 2003) documented the scope, interest, and potential for nurseries to serve tribal needs. In a related need, tribes also requested native plant propagation literature and guidelines including cultural, medicinal, and spiritual plants. This resulted in the Nursery Manual for Native Plants (Dumroese et al., 2009). Both the Needs Assessment and the Nursery Manual still provide critical information concerning tribal nursery activities and technical needs, though funding constraints and limited personnel have delayed their updating, a reoccurring theme among tribal nurseries.

The 2003 assessment included responses from 68 tribes, 7 tribal colleges, and 2 nonprofits. Fifty-two respondents (86%) requested further nursery and restoration training and 38 percent requested environmental education information and lesson plans for their schools. Twenty-seven (35%) of the tribes and tribal colleges had existing nurseries. Most respondents were from small nurseries, ranging from outdoor planting beds for basket materials to small, prefabricated greenhouses. Twenty-four (31%) did not have a nursery but wanted to start one. Most small, existing nurseries wanted to expand the scope of their projects. Ongoing issues identified in the assessment still resonate today:

- Little if any permanent funding for INC activities.
- Lack of availability of some native plants that have high cultural value.
- Loss of traditional ecological knowledge.
- Fragmentation and degradation of important habitats.
- Unemployment and lack of new natural resource professionals to work on reservations.
- Poor relationships among federal, state, and tribal governments.

TRADITIONAL ECOLOGICAL KNOWLEDGE

Although viewpoints on land use, conservation, and management differ widely among and within tribes, to native peoples the land represents their sense of place; a physical, cultural, and spiritual connection which relates to the traditional use of land for subsistence, livelihoods, and well-being (e.g., Hornborg, 2006; Viveiros de Castro, 2004). Dr. Christina Eisenberg of Oregon State University described traditional ecological knowledge as the indigenous wisdom which native people have successfully used to manage natural systems. Such knowledge and practices are passed as narrative histories from one generation to the next to enhance ecosystem management for stability and productivity for food, medicine, and ceremonial items. This knowledge was disregarded by white settlers, who introduced destructive land-management practices, the effects of which persist today (Eisenberg et al., 2019).

---

[3] Indian Nations Conservation Alliance, see https://inca-tcd.org/about-tcds/ (accessed October 23, 2022).
[4] See https://www.rngr.net under Tribal (accessed October 23, 2022).

Historically, research on tribal land has rarely included tribal perspectives even though they often differ from those of non-tribal scientists (Dockry et al., 2022). The integration of tribal viewpoints and traditional knowledge into land management and research (see Chapter 8) is an emerging practice that will advance the capacity of tribal land management to improve ecological restoration and ecosystem function.

INNOVATIVE TRIBAL PLANT PROGRAMS

The Diné Native Plant Program

The committee learned of a 2017–2018 feasibility study conducted to assess nursery interest and needs in the Navajo Nation. One questionnaire was developed for Navajo Agencies and organizations (tribal, federal, and state agencies, and nonprofits working on Navajo land), and one for Navajo community members. As a result, a comprehensive report and plan for native plant needs was developed (Mike et al., 2018). Among agency responses, 67 percent used native plant materials. Common uses included restoration (68%), education (61%), and range rehabilitation (35%). Eighty percent expected a growing need for local plants in the next 5–10 years. Among Navajo communities, 95 percent of respondents reported using native plant materials, most commonly for food (68%), cultural/ceremonial (63%), and medicine (58%). There was an interest in learning about traditional uses of plants (69%) and in partnering with a native plant program to grow and utilize native plants.

The Diné Native Plants Program (DNPP) was established in response to the documented interest and needs of the Navajo Nation. It is designed to meet the needs of both the agency/organizational groups and the interests of community members. For agencies, the focus is on collection of workhorse species for ecological restoration, and for communities, to provide members access to culturally important species. There is no base funding, so grants are essential. Current support for the DNPP comes from the Navajo Nation Government, BIA, Bureau of Land Management (BLM),[5] nongovernmental organizations, and a private foundation.

The Fort Belknap Native Seed and Restoration Program

In fall 2019, the Fort Belknap Indian Community launched a 5-year partnership with BLM and the Society for Ecological Restoration (SER) to implement the Seeds of Success (SOS) Native Seed and Grassland Restoration Program (Eisenburg, 2021). The program focuses on the role of traditional ecological knowledge in ecological restoration, working in close partnership with the Fort Belknap Indian Community on BLM and adjacent tribal lands in Montana's Northern Great Plains. The program applies seed collection in compliance with BLM's SOS program, BLM Assessment, Inventorying, and Monitoring protocols, and ecological restoration based on the SER International Principles and Standards for the Practice of Ecological Restoration.

Tribal Native Plant Materials Program Development Plan for the Confederated Tribes of Grand Ronde

The Institute for Applied Ecology partnered with Confederated Tribes of Grand Ronde (CTGR), the Oregon Parks and Recreation Department, the Oregon Department of Fish and Wildlife, the City of Corvallis, and the NRCS to obtain an Oregon Watershed Enhancement Board restoration grant, which was implemented from 2016 to 2019. The resulting "Plants for People" project focused on utilizing culturally significant plants applying traditional ecological knowledge to restoration. The Institute for Applied Ecology and CTGR created a development plan for an expanded tribal native plant materials program. Continued restoration at Herbert Farm and Natural Area near Corvallis Oregon and the Champoeg State Park adjacent to the Willamette River includes invasive weed control and prescribed burns prior to planting native seed. Prairie and riparian monitoring and photo points are tracking the establishment of native vegetation at these sites.[6]

---

[5] The sentence was modified to add BLM as a supporting organization of the DNPP.

[6] See https://appliedeco.org/report/plants-for-people-bringing-traditional-ecological-knowledge-to-restoration-2018-post-implementation-status-and-plant-establishment-report/ (accessed October 26, 2022).

DISCUSSION

There is undoubtedly much more to uncover about the activities of the tribal nations with respect to native seed that was beyond the ability of the committee to reach, and a deeper exploration of these activities would shed light on seed needs and the tribes' ecological understanding of the landscape. In addition to the insights the tribes could bring to the federal land-management agencies, it is hoped that tribal nations would benefit from greater partnership and support that could contribute to their land-management objectives.

CONCLUSIONS

Conclusion 5-1: *The historic complexity of land-management issues and interactions with federal and state governments and former Bureau of Indian Affairs policies associated with cultural assimilation have constrained all aspects of tribal land management including native plant needs and capacities.*

Conclusion 5-2: *Tribal nurseries are producing plant materials for native plant programs, but the capacity of current nurseries, and interest of many tribes to establish nurseries, is resource limited relative to the needs for native plants among tribes.*

Conclusion 5-3: *Historically, tribal nations have rarely had their perspectives fully integrated into research projects conducted by non-tribal scientists. This has led to mistrust and a dampening of innovative approaches to restoration and integration of traditional ecological knowledge with western science.*

REFERENCES

Dockry, M.J., S.J. Hoagland, A.D. Leighton, J.R. Durglo, and A. Pradhananga. 2022. An assessment of American Indian forestry research, information needs, and priorities. Journal of Forestry 1–15. https://doi.org/10.1093/jofore/fvac030.

Dumroese, R.K., T. Luna, and T.D. Landis, eds. 2009. Nursery Manual for Native Plants: A Guide for Tribal Nurseries - Volume 1: Nursery Management. Agriculture Handbook 730. Washington, DC: US Department of Agriculture, US Forest Service.

Eisenberg, C. 2021. Seeds of restoration success: Fort Belknap Indian Community/BLM/SER Native Seed and Grassland Restoration Program. https://ser-insr.org/news/2021/3/13/seeds-of-restoration-success-fort-belknap-indian-community blmser-native-seed-and-grassland-restoration-program (accessed September 22, 2022).

Eisenberg, C., C. Anderson, A. Collingwood, R. Sissons, C. Dunn, G. Meigs, D. Hibbs, S. Murphy, S. Kuiper, J. SpearChief-Morris, L. Bear, B. Johnston, and C. Edson. 2019. Out of the ashes: Ecological resilience to extreme wildfire, prescribed burns, and indigenous burning in ecosystems. Frontiers in Ecology and Evolution 7:436. doi: 10.3389/fevo.2019.00436.

Hornborg, A. 2006. Animism, fetishism, and objectivism as strategies for knowing (or not knowing) the world. Ethnos 71(1):21–32.

Luna, T., D. Thomas, T.D. Landis, and J. Pinto. 2003. Tribal Nursery Needs Assessment. USDA Forest Service, State and Private Forestry: Reforestation, Nurseries, and Genetic Resources (RNGR), Southern Research Station. https://rngr.net/inc/Tribal%20Nursery%20Needs%20Assessment.pdf (accessed February 10, 2023).

Mike, J., K. Jensen, and N. Talkington. 2018. Diné Native Plants Program: Navajo Nation Native Plant Needs and Feasibility Assessment. Navajo Fish and Wildlife. https://www.nndfw.org/nnhp/docs_reps/NNPP_Feasibility_Assessment_Report.pdf (accessed February 10, 2023).

Murray, M. 2021. Tribal Land and Ownership Statuses: Overview and Selected Issues for Congress. Congressional Research Service Report R46647. https://crsreports.congress.gov/ (accessed February 10, 2023).

Reyes-García, V., Á. Fernández-Llamazares, P. McElwee, Z. Molnár, K. Öllerer, S.J. Wilson, and E.S. Brondizio. 2019. The contributions of Indigenous Peoples and local communities to ecological restoration. Restoration Ecology 27(1):3–8.

Viveiros de Castro, E. 2004. Exchanging perspectives: The transformation of objects into subjects in Amerindian ontologies. Common Knowledge 10(3):463–484.

6

Cooperative Partnerships for Native Seed Development, Supply, and Usage

Because locally adapted native seed can be difficult to find in commercial markets, public and private partnerships have arisen to develop local and regional native seed supplies for restoration, remediation, and other applications, including to provide stock seed to commercial growers. The committee learned of several notable activities taking place at municipal, state, and regional levels, which are described in this chapter. These include state and municipal-level programs, regional programs for seed development, national partnerships for strategies and seed collections, and partnerships for more effective usage of native seed.

STATE AND MUNICIPAL-LEVEL PROGRAMS

The longest-lived of these projects is Iowa Tallgrass Prairie Center, operating in some form for more than 30 years. The newest, the Nevada Seed Strategy, was established in 2020. Some of the partnerships are focused on collection and seed banking, others include plant development, while still others are research-oriented activities to better understand seed zones, soil, and genetics, technologies to improve seed survival, and tools to anticipate how climate change will affect plant communities. Many local partnerships have individual ties to different federal agencies which also collaborate with each other in major regional efforts. They offer models of the building blocks needed for a collective nationwide effort for native seed and plant development and restoration.

Iowa Tallgrass Prairie Center[1]

The Native Roadside Vegetation Center, later renamed the Iowa Tallgrass Prairie Center, was established at the University of Northern Iowa in the late 1970s. The aim of the Center is to restore resilient diverse tallgrass prairies, which have virtually disappeared from Iowa (99.9% lost). The remaining stands of tallgrass prairie are mostly located along gravel roadsides in rural areas of the state, which combined equal about 1.73 million acres. The Center has worked to develop a supply of diverse, local natives to bring back these once ubiquitous landscapes.

According to Laura Jackson, the Center Director who spoke to the committee, it was the Integrated Roadside Vegetation Management (IRVM) legislation passed by the Iowa legislature in 1987 that was a turning point for the Center. State Code 314.22 directed the state to adopt roadside management for state and federal roads and

---

[1] See https://tallgrassprairiecenter.org/ (accessed December 1, 2022).

emphasized the establishment of long-lived vegetation matched to the unique environment found on roadsides, with an emphasis on native plant species. The Iowa Department of Transportation (IDOT) was assigned responsibilities for roadside management, and county participation was encouraged. Specific instructions were given related to mowing, spot spraying, prescribed burning for brush control, and the planting of diverse perennial vegetation. At the time the law was passed, the only plant materials available were western plant cultivars. The Living Roadway Trust Fund, established under State Code 314.22, used tax revenue to support a competitive grants program that provides support for the IRVM work at the county, municipal, and state level and the plant materials development work of the Iowa Tallgrass Prairie Center. The Center now helps counties in adopting and implementing voluntary Roadside Management Plans for their roads by developing seeds and other plant materials through an annual grant from the Federal Highway Administration which enables their free distribution to county roadside projects.

The Center developed a map with three seed zones, has accessioned 100 species through collections, and developed 180 ecotypes (called Iowa Ecotypes) of 88 species. Of those, 155 ecotypes of 82 species have been provided to growers with a total of 531 distributions. Species over time have expanded from warm season to cool season grasses, forbs, and legumes. More recently sedges and shrubs were added to expand functional group diversity. About 1,000 acres of county roadsides are planted annually with seeds from the Center. Iowa Ecotype seed is certified as Source-Identified (see Box 8-2 in Chapter 8 for more information on certified seed), which receives a preference in bids submitted to the IDOT for its roadside maintenance projects. The Center's work has also stimulated the development of similar plant materials by the private sector.

Jackson told the committee that the lessons learned are that public support is needed to help decision makers understand and endorse change. Seed quality, communication, monitoring, adequate funding, and maintaining good records are critical to success. She added that applied research is needed to refine methods and tools for practitioners and managers.

Texas Native Seeds Program[2]

The Texas Native Seeds Program is a not-for-profit native plant seed development program that began as South Texas Natives in 2001, when the revegetation activities associated with the construction of the Interstate 69 highway project attracted the attention of local landowners concerned about non-native invasive plants. The Texas Department of Transportation provided a grant to establish the program, which has since been supported by public and private funders. Keith Pawelek, Associate Director of Texas Native Seeds, told the committee that the program collects, develops, and produces locally adapted seeds of native grasses, forbs, legumes, and woody plants for different regions in Texas. The seeds are released as natural-track certified, Selected (S) (see Box 8-2 in Chapter 8) Texas Native Germplasm and offered to commercial suppliers for large-scale increase, sometimes in blends of selected accessions to broaden the diversity and utility of the seed for locations in different ecoregions or with different soil types. Texas Native Seeds has supported the commercialization of 17 native seed lines and now has projects to provide regionally adapted seed to six different regions in Texas. The Natural Resources Conservation Service (NRCS) E. "Kika" De La Garza Plant Materials Center has been a partner in the Texas Native Seeds Program, along with the Texas AgriLife Research Station, representing a partnership of a nonprofit organization, a state university, the federal government, and the private sector.

Nevada Native Seed Strategy[3]

The Nevada Seed Strategy is a product of the Nevada Native Seed Partnership,[4] whose initial members include the Bureau of Land Management (BLM), US Forest Service (USFS), US Fish and Wildlife Service (USFWS), US Geological Survey (USGS), NRCS, University of Nevada, Reno, Nevada Department of Agriculture, Nevada

---

[2] See https://www.ckwri.tamuk.edu/research-programs/texas-native-seeds-program-tns (accessed February 10, 2023).

[3] See https://agri.nv.gov/uploadedFiles/agrinvgov/Content/Plant/Seed_Certification/FINALStrategy_with%20memo_4_24_20_small.pdf (accessed February 10, 2023).

[4] See https://www.nature.org/en-us/about-us/where-we-work/united-states/nevada/stories-in-nevada/native-seed-restoration/ (accessed February 10, 2023).

Department of Wildlife, Great Basin Institute, the Nature Conservancy, and the Walker Basin Conservancy. Modeled on the National Seed Strategy, its overall aim is "to increase the quantity and quality of [native] seed available for large-scale rehabilitation, reclamation, and restoration (i.e., post-fire seeding) as well as for smaller-scale projects (i.e., wildlife corridors)" (NNSP, 2020, p. 4). The group was formed in 2017, and the first iteration of the Strategy was released in April 2020, putting forward four goals:

1. Identify seed needs, and ensure reliable availability of genetically appropriate seed.
2. Identify research needs to improve technology for native seed production and ecosystem restoration.
3. Develop and implement tools that enable managers and producers to make decisions about collecting, increasing, and using genetically appropriate seed.
4. Develop and implement strategies for internal and external communications.

Each of these goals is being pursued through activities related to several subgoals and objectives that engage different stakeholders and partners. Each year the Nevada Department of Agriculture hosts a Nevada Native Seed Forum to bring producers, research, and land managers and other members of the Nevada Native Seed Partnership together to discuss restoration goals, technologies, and native seed production, fostering both unity and flexibility in the collective approach to the conservation of Nevada's natural heritage. Members work together to identify the most promising seeds for restoration (e.g., Leger et al., 2021), and many of these collections are being agriculturally produced for use in restoration, though seeds are being grown by established seed growers outside the state. To further support the seed industry in Nevada, in 2022, the group began a foundation seed program, providing free seeds of desirable collections to qualified growers within Nevada.[5] Further, partners have multiple field experiments ongoing across the state, designed to test the performance of locally collected seeds in restoration settings.

Utah Great Basin Research Center[6]

The Utah Division of Wildlife Resources (DWR) Great Basin Research Center and Seed Warehouse (GBRC) in Ephraim, Utah, has operated as the technical and logistical hub for revegetation efforts in Utah since 2004. The GBRC's seed warehouse, with capacity for ambient storage of 1.2 million pounds of native and non-native seed and cold (34 degrees F) humidity-controlled (5%) space for 150,000 pounds, makes it one of the largest climate-conditioned restoration seed warehouses in the United States. Utah's Watershed Restoration Initiative[7] (WRI) provides programmatic and administrative support for the GBRC, which purchases seed from wild-seed collectors and agricultural growers in the western United States through a consolidated seed buy of the DWR, and, along with some of its own produced seed, has been building a seed supply for the WRI's proactive restoration efforts that amounts to about 100,000 acres annually as well as fluctuating reactive wildfire revegetation demands. The GBRC coordinates the contracting and implementation of seeding projects, as part of an effort to maintain a strong private industry in Utah and surrounding states. BLM and USFS store limited quantities of seed at the warehouse, and the GBRC also makes seed available to the agencies for purchase through the DWR (Kevin Gunnell, personal communication).

Greenbelt Native Plant Center and Mid-Atlantic Regional Seed Bank[8]

The Greenbelt Native Plant Center (GNPC) is the municipal native plant nursery of New York City's (NYC's) Department of Parks and Recreation and is the nation's largest and longest running city-operated native plant nursery. Since the early 1990s, the Center has provided over 13 million live plants of over 500 species, produced from local plant populations in support of restoration and management of the city's parks and other valuable

---

[5] See https://agri.nv.gov/uploadedFiles/agrinvgov/Content/Plant/Seed_Certification/Foundation%20Seed%20Program%20Application.pdf (accessed February 10, 2023).

[6] See https://cpnpp-natureserve.hub.arcgis.com/documents/Natureserve::utah-division-of-wildlife-resources-great-basin-research-center/about (accessed February 10, 2023).

[7] See https://wri.utah.gov/wri/ (accessed February 10, 2023).

[8] See https://www.nycgovparks.org/greening/greenbelt-native-plant-center and http://www.marsb.org/ (accessed February 10, 2023).

natural areas. Materials from the GNPC have been used in projects staged by multiple city agencies as well as on state and federal projects in the city. When Hurricane Sandy destroyed critical coastal habitat from Maine to the Carolinas in 2012, and including areas of New York City itself, the Center, as a Seeds of Success partner, was able to provide seed and plant material for regional restoration projects.

Ed Toth, a member of the study committee and former director of the GNPC, notes that NYC, not unlike some federal and state agencies, does not employ policies for advanced planning and the City does not pre-fund production of those materials to ensure a supply at the time projects are staged, leaving purchase from the seed supply available "on-hand." Other budgeting and procurement rules and practices, perhaps unique to NYC, prevented fuller use of the facility, despite the tremendous investment the City has made in the nursery.

The GNPC banks about 2,000 collections to meet the production needs of the nursery. These must be replenished on an ongoing basis. Over the last 20 years, almost 13,000 collections have been made toward that end. The City of New York has 27 extant ecosystems and the GNPC has played a role in reintroducing common native species to those habitat niches. The Center's expertise in the production of two-thirds of the extant flora of NYC guarantees that a high level of biodiversity is built into projects. Commercial producers, bound by market constraints and realities, most often are unable to meet these biodiversity needs. This points to the importance of public-sector production of native plant materials as part of the supply chain, in concert with the commercial sector.

The GNPC created the Mid-Atlantic Regional Seed Bank (MARSB)[9] as a programmatic extension of its seed banking resources out into the larger region. To date, the regional seed bank contains 750 accessions and cooperatively banks seed for other native seed programs in the mid-Atlantic region. It has collected seed from the region for BLM's Seeds of Success program, USFS, National Park Service, US National Arboretum, and New York State Department of Parks. MARSB sees itself as a natural partner to any future seed networking efforts in the mid-Atlantic.

Other State-Level Partnerships

The committee learned of other state-level or nonfederal regional partnerships that have formed or are forming. These include the Northeast Seed and Plant Supply Chain Network;[10] the Willamette Valley Native Plant Partnership;[11] the Coastal Native Plant Partnership;[12] the Southwest Seed Partnership;[13] Rhody Native (Reseeding Rhode Island);[14] and the California Native Seed Supply Collaborative.[15]

REGIONAL PROGRAMS FOR NATIVE SEED DEVELOPMENT

Pacific Northwest Region

The Pacific Northwest Region of the USFS (Region 6 of the service's nine regions) is situated across Oregon and Washington and includes 17 national forests, a national grassland, and other scenic lands.[16] Within the region, the Service has developed programs to address all aspects of the seed supply chain. This includes research, seed needs projections under climate change, the refinement of the Seed Selection Tool for decision making in reforestation, as well as the collection, production, cleaning, and storage of seeds of native grasses, forbs, conifers, willows, and other tree species. The program also encompasses infrastructure development, partnerships, funding, and training.

The aim has been to integrate projects among disciplines, with botanists writing plans and coordinating efforts. Projecting seed needs for unplanned disturbances is done by looking at seeding records, fire risk modeling,

---

[9] See http://www/marsb.org (accessed February 10, 2023).

[10] See https://ecohealthglobal.org/network-sites/seed-plant-supply-chain-program/ (accessed February 10, 2023).

[11] See https://www.roguenativeplants.org/willamette-valley-native-plant-materials-partnership-strategic-plan-2013–2017/ (accessed February 10, 2023).

[12] See https://appliedeco.org/restoration/native-seed-partnership/coastal-native-seed-partnership/ (accessed February 10, 2023).

[13] See https://southwestseedpartnership.org/ (accessed February 10, 2023).

[14] See https://riwps.org/reseeding-ri/ (accessed February 10, 2023).

[15] See https://canativeseedcollaborative.wordpress.com/ (accessed February 10, 2023).

[16] See https://www.fs.usda.gov/main/r6/learning/nature-science (accessed February 10, 2023).

and use of other tools. For major seed zones, workhorse species have been selected that have broad ecological applicability, establish well on disturbed sites, are easy to collect and propagate, reliably produce seed, and can be stored for reasonable periods.

Home to the USFS J. Herbert Stone Nursery, the Clarno Propagation Center, the Dorena Genetic Resources Center, the Bend Seed Extractory, and in conjunction with private nurseries and growers, Region 6 has the infrastructure to provide plant materials and restoration advice to its National Forest managers as well as to other federal, state, tribal, and county partners in the region, including Canada. With the passage of the REPLANT Act of 2021,[17] Region 6 will have the resources to do much more restoration work, which is needed to address post-wildfire needs, if it can obtain the personnel needed to carry it out.

Colorado Plateau Native Plant Program[18]

The Colorado Plateau, lying at the intersection of Colorado, Utah, Arizona, and New Mexico, is an arid, challenging environment for plant restoration. BLM established the Colorado Plateau Native Plant Program (CPNPP) in 2007 in partnership with the USFS, Northern Arizona University, and state wildlife agencies in the region. In 2010, an interagency agreement established the USGS Southwest Biological Science Center in Flagstaff, Arizona, as the lead scientific agency for plant material research in the program. There are numerous partners involved in a wide range of activities in the program, including federal, state, and local government agencies; tribal nations; nongovernmental conservation organizations; university researchers and curators; commercial plant materials industry (seed and seedling growers and sellers); and seed testing and certification entities.

The lead scientific agency for plant materials research for the Colorado Plateau project is the USGS. As the science agency for the Department of the Interior, the USGS scientist plays an active role in uncovering the interactions of plant genetics and biophysical aspects of ecological systems, in the context, for example, of the need to identify native plant materials for restoration in the dryland environment of the Colorado Plateau Native Plant Program (Massatti et al., 2022). (See Box 8-1 for a brief description of the research activities of the USGS for the CPNPP.)

Great Basin Native Plant Project[19]

The Great Basin Native Plant Project (GBNPP) was established in 2002 as a cooperative program of the BLM Plant Conservation and Restoration Program and the Grassland, Shrubland, and Desert Ecosystem Research Program of the USFS Rocky Mountain Research Station. Its initiation was part of an interagency response to a 2000 congressional directive asking for a plan to develop a native seed supply for emergency stabilization and restoration. The Great Basin is an expansive swath of land that encompasses large portions of Nevada and Utah, and pieces of California, Idaho, Wyoming, and Oregon. It is called the Great Basin because it is hydrographically contained, with no drainage to the oceans, and ecologically diverse.

The goal of the GBNPP is to develop useful information for land managers faced with decisions about the selection of plant materials for vegetation restoration. The project website boasts of "more than 30 major cooperators in 9 states" working on wide variety of native plant research, development, and restoration projects. At any given time, a dozen or more projects are ongoing. These include understanding the genetic diversity of Great Basin plant species, looking at seed zones considering climate change, investigating cultivation practices for increasing Great Basin seed, species interactions, restoration strategies and equipment, and others. The research activities in the Project have produced a great number of published studies that document such things as variation in local adaptation of Great Basin natives (Baughman et al., 2019); climate niches among forbs in the sagebrush steppe

---

[17] The REPLANT (Repairing Existing Public Land by Adding Necessary Trees) Act will support the reforestation of 4.1 million acres over the next year. It is an element of the Bipartisan Infrastructure Law (Infrastructure Investment and Jobs Act) that became law in 2021.

[18] See https://cpnpp-natureserve.hub.arcgis.com/ (accessed February 10, 2023).

[19] See http://www.greatbasinnpp.org/ (accessed February 10, 2023).

(Barga et al., 2018); and using the genetic structure and history of blue-bunch wheatgrass populations as foundational knowledge for developing plant materials for restoration (Massatti et al., 2018).

Mojave Desert Native Plant Program[20]

The Mojave Desert ecoregion is situated across Nevada, California, and parts of Arizona and Utah in some of the most diverse and arid landscapes in the nation. The region is challenged by the impacts of invasive species, recreational activities, and a megadrought—the driest decade in 1,200 years (Williams et al., 2022). Initiated by BLM in 2017, it has developed into a partnership with the USGS, the Rancho Santa Ana Botanic Garden, NRCS Tucson Plant Materials Center, Utah Division of Wildlife Resources, Texas State University, Victor Valley College, and others, to collect and bank seeds, develop germplasm releases for commercialization, conduct landscape genetic studies, and develop empirical seed transfer zones. The program also conducts research on restoration techniques and strategies. It has identified priority plant species for collection (Esque et al., 2021), and developed a "Mojave seed menu" (Shyrock et al., 2022) as a decision-making tool for restorationists, incorporating climate change projections.

NATIONAL PARTNERSHIPS

Seeds of Success

The BLM Seeds of Success (SOS) program, which collects and banks seed to preserve the genetic diversity of native plant populations for future use in restoration, was developed in 2001 as a partnership with the Kew Gardens Millennium Seed Bank project. It was also one of the first steps taken that addressed the congressional mandate of 2002 for the development of a supply of native seeds for emergency stabilization and restoration. Federal partners include the USFS Bend Seed Extractory, the US Department of Agriculture's (USDA) Agricultural Research Service, and the USFWS. In 2008, BLM signed a Memorandum of Understanding (MOU) with six nonfederal partners to perform collections under a common protocol and to store seed for research and working collections. The agreement formalized SOS as the national native seed collection program for conservation and restoration across the United States, with BLM's Plant Conservation and Restoration Program as the lead. The MOU states, in part:

> While the Federal land managing agencies purchase large quantities of native seed to meet their needs, the demand for native seed is also felt by state and local land managing agencies and other non-governmental organizations engaged in land management and conservation. A reliable, sustainable, and ecologically appropriate source of seeds is needed to meet all these needs and can best be met within a national framework for management and conservation of the nation's seed resources. Such a framework should seek to equitably address the demand for regional and local seed by assuring a ready supply through appropriate planning, coordinating, collection and storage, while protecting the resources. It is vitally important for the BLM to work with partners such as the Botanic Gardens, to achieve nationwide restoration goals.[21]

The SOS program has continued to add nonfederal partners over the years, and by the end of 2021 had made over 27,000 accessions of 5,800 unique taxa. From 2014 through 2016, BLM funded an eastern collection program, Seeds of Success East. This effort successfully made 2,124 seed collections and has made that seed available to 27 federal, state, and municipal agencies for post-Hurricane Sandy recovery and resiliency projects.

Most recently the SOS has supported two new partnerships, one between BLM Montana and the Fort Belknap Indian Community to use SOS seed collection protocols and traditional ecological knowledge for collecting and restoring tribal lands, and a second, to establish a Seeds of Success Southeast program. With BLM support, the

---

[20] See https://gis.blm.gov/DETODownload/Docs/The_Mojave_Desert_Native_Plant_Program.pdf (accessed February 10, 2023).

[21] See https://www.blm.gov/sites/blm.gov/files/program_nativeplants_collection_homepage_SOS%20MOU.pdf, p.2 (accessed February 10, 2023).

USFWS Partners for Fish and Wildlife Program and the Southeastern Grasslands Initiative will coordinate with the national SOS to develop a collection strategy over 10 states in the southeastern United States.

Agricultural Research Service and SOS

The Agricultural Research Service (ARS) provides research in many areas, and presentations by Brian Irish and Stephanie Greene provided the Committee with information about two programs that are particularly germane to this report.[22] The National Plant Germplasm System (NPGS) has 25 gene banks, many partnering with land grant universities, with the goal of conserving plant germplasm for agricultural and other uses. About 600,000 accessions are in storage; most are available in small quantities to private and public users for research, plant breeding, evaluation, and education. Currently, the NPGS provides long-term storage for small accessions of collections from the Seeds of Success program. The Germplasm Resources Information Network includes data for about 16,000 species stored at the NPGS.

Irish and Green expressed a need for a dedicated curatorial program for natives within the NPGS because most have no curatorial support and often lack production and germination protocols, making their conservation and curation difficult. Such a program might require resources of $2 million annually for a subset of natives; native species that are wild-crop relatives would have greater relevance to a larger group of stakeholders and would be more attractive targets for curation at the NPGS.

Seed Cleaning for SOS

The USFS Bend Seed Extractory in Bend, Oregon, currently cleans all the seed collected by the western SOS seed collection program with fees paid by BLM through an agreement with the USFS and cleans seed for other government clients on a fee-based system. However, it is not of unlimited capacity, and maybe increasingly focused on cleaning conifer seed as forest restoration efforts under the REPLANT Act begin to increase. It could serve, at least in part, as a useful template for establishing cooperative seed-cleaning efforts across the United States.

There are limited examples of cooperative efforts to provide seed cleaning and seed warehousing. Yet they are needed for a national supply chain based on ecoregional seed of many different species. Many seed producers of sufficient size can clean their own seed crops. However, the more species that are grown, especially if they are forbs, the more specialized cleaning equipment is needed. This can be cost prohibitive, especially to smaller producers, and creates a significant disincentive for such producers to consider entering the restrictive markets that ecoregional seed production imposes.

Seed Warehousing and SOS

Apart from the long-term frozen storage of SOS accessions in the seed banks of the ARS National Plant Germplasm System, seed collected through SOS is stored in seed banks of partner organizations. Seed banking of wild seed, both for conservation purposes and to meet restoration and other land-management needs, can be effectively managed at regional seed banks that have appropriate storage capacity for wild collections and appropriate technical staff to curate wild collections and ensure their long-term quality and viability.

Seed warehousing is an extension of seed banking, but particularly focused on large quantities of increased seed typically held for relatively short periods of time because seeds are short-lived in ambient warehouse conditions. It is most relevant to the other end of the supply chain, where an adequate supply of increased seed must be maintained to ensure a timely source for end users. This is necessitated by the long timeline for development and production of increased seed, which in turn dictates the need for timely planning, production, and storage well in advance of need. Only a handful of federal native seed warehouses exist, and all are in the arid West.

---

[22] See https://www.plantconservationalliance.org/sites/default/files/PCA%20NPGS%20-%20Native%20PGRs%20%2805-10-2021%29_1. pdf and https://www.youtube.com/watch?v=BJVSwV3rym4 (accessed February 10, 2023).

Similarly, some states in the arid West maintain native seed warehouses to meet their needs. If the supply and demand for ecoregional native seed increases, as is anticipated, seed warehousing space will become an issue.

The essential storage requirements for seed are low temperatures and low levels of relative humidity. Although in the arid West seed warehouses naturally experience low relative humidity for most of the year, temperature and humidity both can vary over the year. Nonetheless, most storage is at ambient temperature without any controls and therefore seed typically does not last more than a few years. Outside of the arid West, both control of relative humidity and cold temperatures are essential for maintenance of seed viability during the extended storage periods that are needed to maintain adequate supplies in advance of need. To date, there are no federal seed warehouses in the East. In all cases, warehouses with dry and cold and/or frozen storage would greatly extend shelf-life of seeds.

Since 2018 BLM has used its Indefinite Delivery Indefinite Quantity to contract for the increase of SOS seed of different species collected from across numerous seed zones. As of 2021, BLM had contracted for over 94,000 pounds of seed expected to arrive between now and 2025, which will be put into the BLM seed warehouses for use by BLM emergency stabilization, wildlife habitat, pollinator, oil and gas, and other programs.

National Seed Strategy

In 2015, the Plant Conservation Alliance, an association of more than 400 nonfederal cooperators and (initially) 12 federal agencies, produced the National Seed Strategy "to foster interagency collaboration to guide the development, availability, and use of seed needed for timely and effective restoration."[23] That document describes itself as a framework that puts forward four broad goals and objectives:

1. Identifying and quantifying seed needs,
2. Conducting research,
3. Developing tools for land managers, and
4. Ensuring communications.

Much like the 2002 plan requested by Congress, it offers a conceptual structure for activities of the agencies. The 4 Goals, divided into 14 Objectives, which are further divided into 51 Actions, are all directed at the overall goal of increasing the supply of "the right seed, in the right place, at the right time." The *National Seed Strategy* is the call to action. In fiscal year (FY) 2021, that call generated a robust response of a combined investment of $271 million and the involvement of 17 federal agencies, 27 tribes, and 481 nonfederal partners in projects contributing to the 163 projects that were under way.

A summary of achievements in FY 2021 includes the collection of 1,434 native species, agricultural production of 1,118 native species, research on 692 species, production of 91,208 pounds of native seed, support for 118 farmers, 20 new facilities, 153 federal and 138 new nonfederal jobs, and the restoration of 30,466 acres (PCA, 2022).

New Funds for the National Seed Strategy

The Bipartisan Infrastructure Law has allocated $200 million ($70 million to the Department of the Interior, $130 million to the Department of Agriculture, the latter for significant increase in tree planting by the USFS) over the next 5 years for the implementation of the Strategy, which may present an opportunity to begin to scale up the efforts currently under way.

PARTNERSHIPS FOR MORE EFFECTIVE SEED USAGE

In addition to buying and using native seed directly, the federal government also affects native seed sales and usage through its support of programs that encourage, fund, and guide native seed usage in the private sector and with state and tribal governments. One of the largest of these is the USDA Conservation Reserve Program (CRP),

---

[23] See https://www.usgs.gov/news/national-seed-strategy-progress-report, p.1 (accessed February 10, 2023).

which covers 21 million acres. Roughly half of that area is planted with native plant species. Other federal agencies provide awards and grants for seed purchases by states, or provide financial and technical assistance, such the USFS Cooperative Forestry program.

USDA Conservation Reserve Program[24]

The CRP is a 35-year-old Farm Bill program of the USDA that incentivizes landowners to place marginal farmland into conservation uses via a 10–15-year contract, which may be renewed. These uses may include improving soil health, water quality, wildlife habitat, carbon sequestration, and pollinator support, although they do not include full ecosystem restoration or ecological integration into existing natural habitats. The program is voluntary, and landowners receive financial incentives and technical assistance. For larger areas, landowners competitively bid for inclusion in the CRP by indicating the conservation practice they will use on the land. For smaller areas, the program is non-competitive and funded on a first come, first served basis.

There are different practices for which landowners can subscribe, some using native plants and some using non-native plant species. In a presentation to the committee, Rich Iovanna of the NRCS and Bryan Pratt, Economic Research Service, estimated that 10 to 13 million acres of the CRP's current 21 million acres are planted with native species, although some non-natives are included in some "native" mixes. The species mix used by a given CRP landowner is specified by a conservation plan prepared by the landowner and approved by the NRCS, following recommendations made by NRCS county and state staff. In competitive proposals submitted by landowners, the inclusion of natives can increase the points given to bidders. The native species that are recommended and used in these mixes are typically based on native seed developed by the NRCS Plant Materials Centers (PMCs). The majority of these natives are commercially viable and are derived from the 575 plant lines developed and maintained by the NRCS-PMCs and released to commercial growers for further increase. The seeds are purchased by landowners from private vendors, assisted by cost-sharing from the CRP.

The influence of CRP on the native seed market was noted by Laura Jackson (Director, Tallgrass Prairie Center, Iowa), who told the committee that the recent initiation of a pollinator-enhancing conservation practice in CRP created shortages of certain native seeds because of their popularity in many plans. Landowners receive about $200/acre reimbursement as a cost-share for adopting this pollinator practice. However, because the seed was not required to be of a local ecotype, there was an influx of less-expensive seed that caused certified seed to be comparatively expensive. Coincident with the arrival of the pollinator seed, growers reduced the use of certification services, which are optional for seed sellers. Certification validates the geographic origins of seed, but the service adds to the cost of seed (see Chapter 8, Box 8-2).

A report from the USDA Economic Research Service (Pratt and Wallender, 2022) concluded that landowners most often bid to install a "premium" native grass mix (which includes forbs), which is four times more expensive than a simpler "base" mix of non-native grasses ($107 versus $25 per acre). The paper observes that the cost of seed is a significant fraction of the total cost of installing the practice and that it is significantly more expensive in the Midwest to upgrade from base to premium ($65 to $112 per acre) than in the Plains and Mountain states ($21 to $43 per acre). The authors calculate that decreasing the cost of the seed by $10 per acre would move 0.6 percent of acreage to the higher-quality practice. Alternatively, they postulate that increasing the cost-share provided to landowners would also incentivize an upgrade to plant a higher-quality mix of natives.

John Englert (NRCS National Program Leader) observed that the diversity and ecosystem functionality of CRP lands, as well as other lands on which NRCS assists private landowners, could be improved if PMCs could devote more resources to developing native materials, including local ecotypes.

USFS Programs that Support Private Land

The USFS manages federal forested lands, but its mission reaches beyond federal lands through its State and Private Forestry outreach to states, tribes, communities, and non-industrial private landowners. Programs in this

---

[24] See https://www.fsa.usda.gov/programs-and-services/conservation-programs/conservation-reserve-program/ (accessed February 10, 2023).

area provide technical and financial assistance to landowners and resource managers to help sustain forest and grasslands, protect communities from wildland fires, and restore fire-adapted ecosystems. In doing this, the federal investment leverages the capacity of its partners to manage state, tribal, and private lands. Examples include Forest Health Protection that provides technical assistance on such things as native and non-native insects, pathogens, and invasive plants. The Cooperative Forestry program provides financial and technical assistance to landowners, communities, and businesses to actively manage and sustain long-term investment in nonfederal forest land.

US Department of Transportation's Federal Highway Administration

The US Department of Transportation (USDOT) allots funds to the state departments of transportation (DOTs) through the Federal Highway Administration (FHWA). Since 1987, the use of native plants in roadside construction and maintenance has been incentivized by FHWA requirements for at least one-fourth of 1 percent of the cost of roadside landscaping to include native wildflowers.[25] Many state DOTs have pursued the expansion of native plantings that support pollinators and the Monarch butterfly.[26] As described in this chapter, the DOTs in both Iowa and Texas have played important roles in assisting efforts to develop local ecotypes of native plants.

DISCUSSION

The committee's investigation into cooperative partnerships further broadened its view of the national capacity to generate a supply of native seeds and revealed their potential in growing the native seed supply. First, federal agencies have developed, engaged in, and supported cooperative partnerships for over 20 years and these partnerships play a central role in helping them to meet their needs for seed. The large, regional native plant materials development and restoration programs initiated in the West by BLM and USFS are broadly cooperative, and involve government at all levels, the tribes, colleges, nongovernmental organizations, and native seed producers. BLM established a national native seed collection program, in partnership with six nonfederal seed banks as the stewards of those collections. And, along with more than 400 nonfederal cooperators, 12 federal agencies have committed to an expansive strategy with 51 Action Steps of the National Seed Strategy, each step a collaboration of different players, collectively aimed at bringing land managers the plant materials and decision tools they need to lead successful ecological restorations.

Second, time and again, independent efforts have arisen in states and regions across the United States in response to the need for native plant materials. New efforts continue to develop, some including federal partners, some without, but all involve cooperators, and several have developed business plans for expansion. These findings suggest that

- Regional cooperative efforts have shown to be successful approaches to tackle the native plant material supply issues.
- Seed resources and seed needs cross jurisdictional lines. With climate change and fragmented, shrinking habitat, it will increasingly be essential to develop cooperative arrangements that ensure that jurisdictional authorities are not barriers to securing a reliable supply of restoration materials for everyone.
- Existing regional efforts can benefit by expanding to include additional federal and nonfederal partners.

Despite these efforts, these partnerships have not yet been able to move the native seed supply chain over the threshold to achieve a level of reliability and predictability that is needed to offer a sustainable supply of a diverse native seed. Continued shortages of seed suggest that the supply chain is underdeveloped, and in some places lacking altogether, when what is required to meet the growing needs is an enterprise that is more robust and at a much larger scale. In the committee's assessment, achieving that goal will not be possible without all the

---

[25] See https://www.environment.fhwa.dot.gov/env_topics/ecosystems/roadside_use/vegmgmt_rdus3_2.aspx (accessed February 10, 2023).

[26] See https://www.environment.fhwa.dot.gov/env_topics/ecosystems/roadside_use/vegmgmt_rdsduse4.aspx (accessed February 10, 2023).

key players that play a central role in the native seeds supply chain, including native seed suppliers. In the next chapter, the committee provides its findings from suppliers, who are central to solving the supply puzzle.

In addition, the committee sees a critical role for leadership at the federal level to complement, extend, and strengthen the work of regional partnerships. Some regions do not yet have seed partnerships, and federal agencies can take the lead in their establishment. Existing and future regional partnerships will function more effectively if there is a coordinated nationwide approach to policies, research, data collection, and information sharing.

CONCLUSIONS

Conclusion 6-1: *There are several examples of successful seed development, seed production, and restoration programs initiated at the state level, sometimes begun with state support and sometimes with federal support.*

Conclusion 6-2: *Over the last 20 years, regional programs initiated by BLM and USFS to meet their native seed needs have included an array of partners from state and local governments, universities, and the private-sector native seed and restoration industries.*

Conclusion 6-3: *The Seeds of Success program should continue as a national effort and expand its collection activities across the United States and increase its cooperative relationships with regional seed banks. The availability of seed supplies for native shrubs will remain reliant on private-sector collectors, as to date, efforts to produce shrubs in seed fields have not proven economical.*

Conclusion 6-4: *An expansion of temperature- and humidity-controlled seed warehouses across all regions of the United States would support the ability to have seed available for when it is required.*

Conclusion 6-5: *The ready availability of cooperative seed-cleaning facilities is important to encourage smaller-scale producers to enter the native seed supply chain, especially in areas where commercial facilities do not exist, or are unwilling to clean native seed.*

Conclusion 6-6: *Although the Conservation Reserve Program is not an ecosystem restoration program, the use of native seeds on private lands contributes to functioning ecosystems and plays a role in supporting the native seed supply. Private landowners seem willing to implement conservation practices with higher ecological values on their land to enhance the competitiveness of their bids for inclusion in the program. Lowering the cost of native seeds to landowners, potentially through a subsidy or premium, would incentivize more landowners to plant native seeds.*

REFERENCES

Barga, S.C., T.E. Dilts, and E.A. Leger. 2018. Contrasting climate niches among co-occurring subdominant forbs of the sagebrush steppe. Diversity and Distributions 24:1291–1307. doi:10.1111/ddi.12764.

Baughman, O.W., A.C. Agneray, M.L. Forister, F.F. Kilkenny, E.K. Espeland, R. Fiegener, M.E. Horning, R.C. Johnson, T.N. Kaye, J. Ott, J.B. St. Clair, and E.A. Leger. 2019. Strong patterns of intraspecific variation and local adaptation in Great Basin plants revealed through a review of 75 years of experiments. Ecology and Evolution 9:6259–6275. https://doi.org/10.1002/ece3.5200.

Esque, T.C., L.A. Defalco, G.L. Tyree, K.K. Drake, K.E. Nussear, and J.S. Wilson. 2021. Priority species lists to restore desert tortoise and pollinator habitats in Mojave Desert Shrublands. Natural Areas Journal 41(2):145–158. https://doi.org/10.3375/043.041.0209.

Leger, E.A., S. Barga, A.C. Agneray, O. Baughman, R. Burton, and M. Williams. 2021. Selecting native plants for restoration using rapid screening for adaptive traits: Methods and outcomes in a Great Basin case study. Restoration Ecology 29(4):e13260.

Massatti, R., H.R. Prendeville, S. Larson, B.A. Richardson, B. Waldron, and F.F. Kilkenny. 2018. Population history provides foundational knowledge for utilizing and developing native plant restoration materials. Evolutionary Applications 11(10):2025–2039.

Massatti R., D.E. Winkler, S. Reed, M. Duniway, S. Munson, and J.B. Bradford. 2022. Supporting the development and use of native plant materials for restoration on the Colorado Plateau (Fiscal Year 2021 Report). Report submitted to the US Department of the Interior Bureau of Land Management on May 5, 2022.

NNSP (Nevada Native Seed Partnership). 2020. Nevada Seed Strategy. https://agri.nv.gov/uploadedFiles/agrinvgov/Content/Plant/Seed_Certification/FINALStrategy_with%20memo_4_24_20_small.pdf (accessed February 10, 2023).

PCA (Plant Conservation Alliance). 2022. National Seed Strategy Progress Report 2015–2020. https://www.blm.gov/sites/blm.gov/files/docs/2021-08/Progress%20Report%2026Jul21.pdf (accessed February 10, 2023).

Pratt, B., and S. Wallander. 2022. Cover Practice Definitions and Incentives in the Conservation Reserve Program. EIB-233, U.S. Department of Agriculture, Economic Research Service.

Shryock, D.F., L.A. DeFalco, and T.C. Esque. 2022. Seed Menus: An integrated decision-support framework for native plant restoration in the Mojave Desert. Ecology and Evolution 12(4):e8805.

Williams, A.P., B.I. Cook, and J.E. Smerdon. 2022. Rapid intensification of the emerging southwestern North American megadrought in 2020–2021. Nature Climate Change 12(3):232–234.

7

Seed Suppliers

INTRODUCTION

This chapter reports on findings of the Committee from its survey of firms that supply native seed and native plants (seedlings and vegetative materials). As with the state agency buyer survey, methodology details appear in Chapter 2.

The supplier survey sought to explore supply-side observations from the Committee's 2020 interim report (see Box 1-2). In particular, the survey explored the research questions that emerge from the interim report's Observations 3, 5, 6, 7, and 8:

- How do timeframe, quantity, and quality affect the production decisions of native seed suppliers?
- What factors affect buyer costs of production for native seed?
- How and to what degree do communication, market and production risk, and the design of contracts with buyers affect what and how much native seed suppliers elect to produce?
- How adequate are suppliers' storage facilities for native seed?
- How do the factors above vary by region of the country (West vs. East) and scale of production?

The chapter begins by characterizing the firms that supply native seed by size, region, product mix, and source of seed genetics. It proceeds to examine the buyers and contract arrangements used, as well as how suppliers anticipate likely demand. Given time lags for seed production and the time and cost of producing locally adapted native seed, the chapter examines how suppliers cope with risky production conditions and uncertainty about which native seed species will have markets at the end of the production cycle. The chapter reports supplier responses on the share of year-end unsold native seed inventories and the adequacy of storage facilities to keep seed in good condition. The chapter closes with suppliers' views on the challenges to expanding the supply of native seeds and native plants.

BUSINESS ACTIVITIES OF FIRMS—PRODUCTS OFFERED

As discussed, the Committee's goal for the supplier survey was to focus on businesses that sell native seed and/or plants, either exclusively or as part of their broader offerings. At the beginning of the survey, respondents were asked several questions about the seed and plants they sell. Among the firms contacted, 81 percent said that

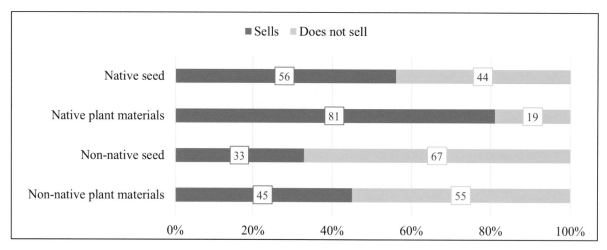

FIGURE 7-1 Types of seed and plants sold by respondents to the supplier survey.

they sell native plants, while 56 percent indicated that they sell native seed, 33 percent sell non-native seed, and 45 percent sell non-native plants (Figure 7-1). Because the intent of the survey was to learn about the experiences of suppliers of native seeds and plants, respondents who said that they do not sell natives of either seed or plants were deemed ineligible and not asked further questions about their business.

Among the eligible respondents, businesses from the West were more likely than those in the East to sell non-native seed (38% vs. 29%), but for other types of seed or plant material, the answers from East and West suppliers were similar.[1] To obtain a sense of the size of the businesses included in the survey, suppliers were asked to provide an estimate (choose from six categories) of their average annual sales and operating revenues during the period 2017–2019. Suppliers with larger sales revenues were more likely to sell seed (native and non-native) than businesses with smaller sales. This may be related to a customer base that includes private landowners that buy native and non-native seed for hay or rangeland. Smaller businesses were more likely to sell native plants than larger businesses (Figure 7-2).

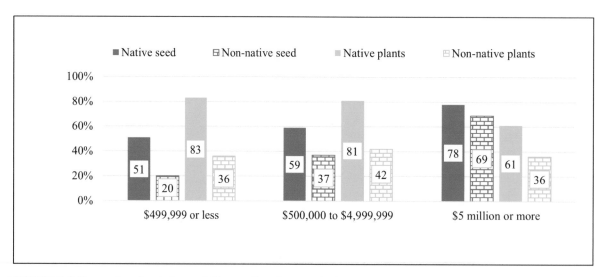

FIGURE 7-2 Types of seed and plants sold by suppliers, by annual sales.

_____

[1] The committee used the same definition of East and West as for the survey of state departments. East includes AL, AR, CT, DE, FL, GA, IA, IL, IN, KS, KY, LA, MA, MD, ME, MI, MN, MO, MS, NC, ND, NE, NH, NJ, NY, OH, OK, PA, RI, SC, SD, TN, TX, VA, VT, WI, WV; West includes AK, AZ, CA, CO, HI, ID, MT, NM, NV, OR, UT, WA, WY.

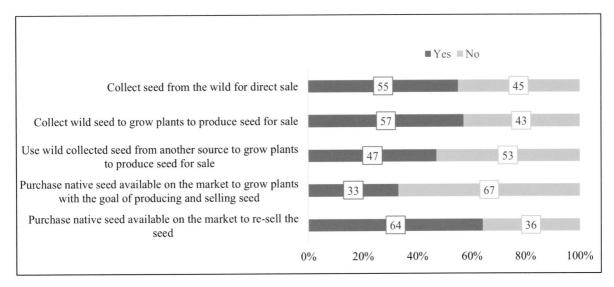

FIGURE 7-3 Supplier's source of native seed.

Suppliers who sold native seed were asked questions about where they obtain the seed they sell (Figure 7-3). More than half sell seed they have collected from the wild, and about two-thirds purchase seed on the market to resell directly. This suggests that many suppliers may be seed consolidators, or that buy seed to expand their range of product offerings. Fifty-seven percent will plant seed they wild collected to grow plants to increase seed for sale, and one-third will use seed they purchased on the market for that purpose. A little less than half will do the same with wild-collected seed obtained from another source, such as a collaborating organization.

Suppliers who said they sold native seed were asked whether they sold several different categories of native seed (Figure 7-4). Approximately two out of three (68%) reported selling "Uncertified seed of claimed wild origin" and two out of three (65%) sell "Certified source-identified seed."

This question was further examined for differences in regions. Around 70 percent of western suppliers and 61 percent of eastern suppliers indicated they sold certified Source-Identified seed. Suppliers from the West were more than twice as likely to sell "Seed types meeting Bureau of Land Management (BLM) requirements" than

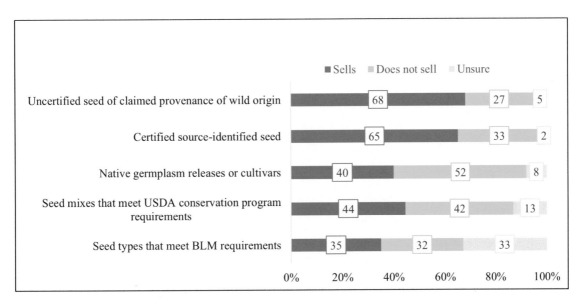

FIGURE 7-4 Types of seed sold by native seed suppliers.

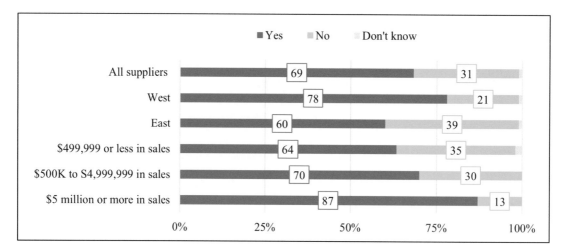

FIGURE 7-5 Sale of native plants or seeds sourced from different locations to match different geographical conditions or seed zones (different ecotypes of the same species), by region and size.

Eastern suppliers (49% vs. 22%). About half (48%) of suppliers from the East and two out of five (40%) of Western suppliers said that they sell "Seed mixes complying with USDA conservation programs."

Native plant and seed suppliers were asked if they sold plants or seeds from different locations, to match different geographic conditions or seed zones, as in different ecotypes of the same species. This question was summarized by region and by sales revenue (Figure 7-5). While nearly 70 percent of suppliers overall offered plants or seed of this nature, they were more likely to be offered by suppliers with the largest sales revenue (87%). Western suppliers were more likely than eastern suppliers (78% vs. 60%) to offer plants or seeds matched to different geographies or seeds zones.

Suppliers were asked the percent of their total native seed sales that were represented by native seed of different plant types (grasses, forbs, shrub, trees, or other). Figure 7-6 shows a profile of suppliers and the concentration of product types as a proportion of their total native seed sales, ranging from 0 percent of their total sales to 100 percent of sales on the x-axis. For example, the first set of bars on the left shows that almost 60 percent (58%) of growers reported that seeds of native trees represented 0 percent of total sales. The set of bars on the far right of the graph shows that seeds of native trees make up 100 percent of the total native seed sales of about 8 percent of suppliers.

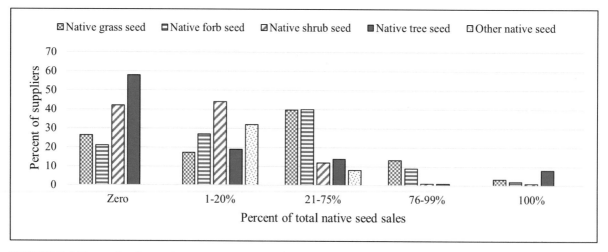

FIGURE 7-6 Types of native seed sold by suppliers in relation to total native seed sales.

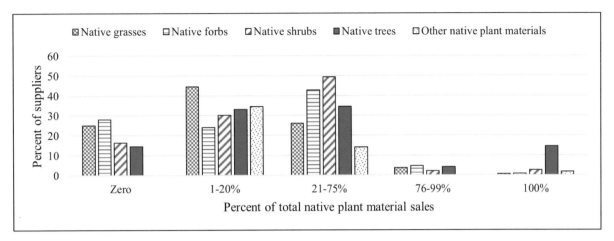

FIGURE 7-7 Types of native plants sold by suppliers in relation to total plant material sales.

In the middle set of bars, 40 percent of suppliers said that seed of native grasses and forbs made up between 21 and 75 percent of total sales. Much lower proportions of suppliers said seeds of native shrubs, trees, or "other" made up between 21 and 75 percent of total sales. The figure suggests that seeds of native shrubs and native trees represented a smaller proportion of total sales, relative to grasses and forbs, for the majority of suppliers in the survey. The means (M) and medians (Mdn) for the supplier responses to the question about what percentage of their total native seed sales were represented by each category were as follows: native grass seed, M=37, Mdn=30; native forbs seed, M=32, Mdn=25; native shrub seed, M=9, Mdn=2; native tree seed, M=17, Mdn=0; other native seed, M=5, Mdn=0.

Similarly, suppliers of native plants (as opposed to native seed) material were asked the percent of their total native plant sales that were represented by each native plant material of different type (native grasses, forbs, shrubs, trees, or other). Figure 7-7 shows a profile of suppliers and the concentration of product types as a proportion of sales (expressed in groups of increasing proportion).

The bars in the middle of the figure show that about half (49%) of the suppliers indicated that native shrub plant material made up 21 to 75 percent of their total plants sales. Forty-three percent and 35 percent of suppliers, respectively, said that native forbs and native trees made up 21 to 75 percent of their total sales. Forbs appear to be sold as much in plant form as seed, given that 40 percent or more suppliers indicate they make up between 21 and 75 percent as a percent of sales of both seeds and plants. The mean and median for the supplier responses to the question were as follows: native grasses, M=18, Mdn=10; native forbs, M=25, Mdn=20; native shrubs, M=25, Mdn=25; native trees, M=35, Mdn=25; other native plants, M=10, Mdn=1.

Business Activities of Suppliers—Other Services

Suppliers were asked in an open-ended question if they sell services and if so, to specify the services they offer. Thirty-six percent of those that answered the question said they provide services. Seed cleaning and installation (site preparation and planting of seeds and trees) were the two most frequently offered services, each mentioned by approximately one in three of those who said that they offered services in addition to selling seed or plants. Other services mentioned included consulting and design, seed collection, ecological restoration, and contracting growing. A few suppliers said that they offer landscaping and weed management.

Business Activities of Suppliers—Clients

The survey asked suppliers to indicate what relative portion of their annual sales were represented by sales to different buyers. Figure 7-8 shows the importance of private contractors to most suppliers—over 60 percent of suppliers said that private contractors represent a major portion of their sales. Approximately one in four

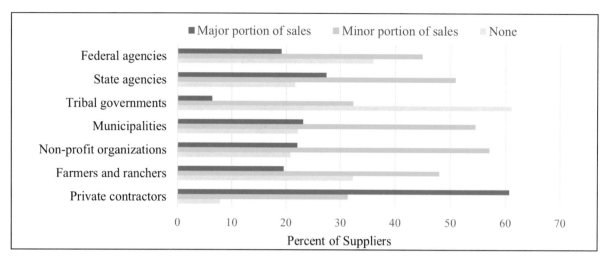

FIGURE 7-8 Portion of annual sales of native seed and plants represented by various customer types.

suppliers said that state agencies were a major portion of sales. About the same fraction of suppliers (23% and 22%, respectively) indicated that municipalities and not-for-profit organizations were a major portion of sales. About 19 percent of suppliers said that farmers and ranchers and federal agencies represented a major portion of sales, and even fewer (7%) said that tribal governments were a major portion of sales. As states, municipalities, tribes, and the federal government often contract out seeding work, the importance of contractors in sales may obscure the true "buyers" of seed for seeding projects.

Business Activities of Suppliers—Communications

Suppliers were asked about possible methods for communicating with potential buyers about what their business offers (Figure 7-9). The great majority of suppliers indicated that a company website (85%) and/or word of

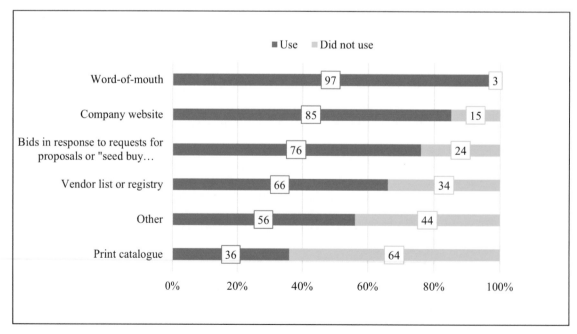

FIGURE 7-9 Suppliers' methods of communicating about what the business offers.

mouth (97%) were used. Three out of four suppliers communicate through their bids to Requests for Proposals and two out of four mentioned vendor registries.

When asked to specify what "other" methods suppliers use to communicate with potential buyers, social media was the most frequent method mentioned, along with emails and newsletters. Some suppliers also mentioned trade shows, seed associations, and native seed groups. A few said that their long history in the business and long-standing customers serve as a means of communication with potential buyers.

Business Activities—Contracting Arrangements

The survey asked suppliers whether they used certain types of contracting arrangements:

- "Bids on consolidated seed buys," which are responses to public procurement requests made by government agencies;
- "Spot market sales of available seed or plants," which are exchanges between buyers and sellers that do not specify any conditions of sale or production;
- "Marketing contracts," which specify the type, price, quantity, and delivery date of seed or plants and guarantee a purchaser for that seed or plants; and guarantee a purchaser for that seed or plant materials; and
- "Production contracts" which share some production costs and/or production risks by providing flexibility in the quantity delivered and/or delivery date.

Figure 7-10a shows that about 23% of all suppliers bid on consolidated seed buys. Each of the other three types of contracting arrangements are used by about half of all suppliers. Thus, slightly less than half as many suppliers bid on consolidated seed buys than use any of the other contracting arrangements. BLM operates the consolidated seed buys to obtain seeds largely to address post-wildfire needs in the West.

Approximately half of the suppliers who provided an answer to the questions about contracting arrangements said that they use more than one type. Utilizing multiple contracting types may reflect differences in buyer and/or supplier preferences. It may also reflect differences in suppliers' revenues and regional markets.

The analyses were also examined by location of the supplier in the United States. Figure7-10b shows that about one in four (24%) suppliers in the West and one in five (21%) suppliers in the East indicated that they bid on consolidated seed buys, which is surprising because the BLM consolidated seed buys typically seek species and ecotypes for use in the West. Suppliers in the West were more likely to use marketing and production contracts

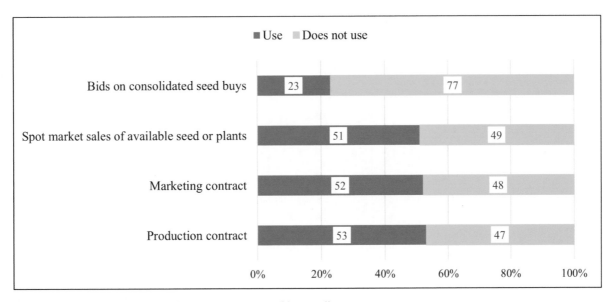

FIGURE 7-10a Types of contracting arrangements used by suppliers.

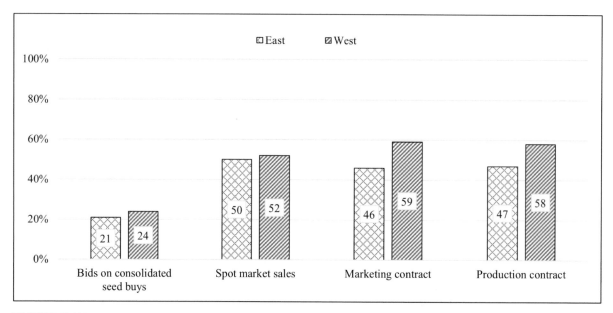

FIGURE 7-10b Types of contracting arrangements used by suppliers by region.

than suppliers in the East. Suppliers in the West appeared slightly less likely to sell via spot market sales than to sell using marketing and production contracts. In contrast, suppliers in the East appeared slightly more likely to sell via spot markets than to sell using marketing and production contracts.

The analyses were next summarized by categories of sales revenue. Figure 7-10c shows that a higher percent (57%) of firms in the category of highest annual sales revenue ($5 million or more) bid on consolidated seed buys versus suppliers with mid-size or smaller revenues (26% and 13%), respectively. The same is true for spot

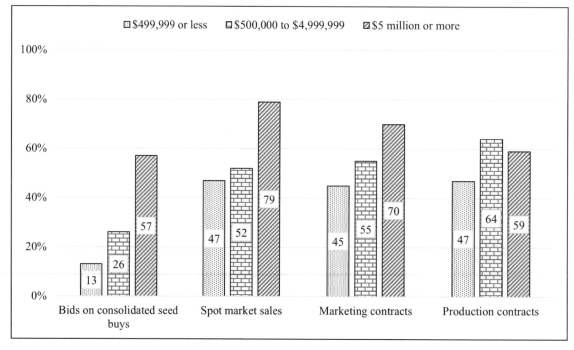

FIGURE 7-10c Types of contracting arrangements used by suppliers, by annual sales.

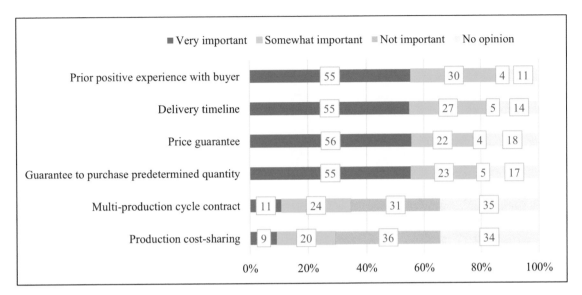

FIGURE 7-11a Importance of contract characteristics to suppliers.

market sales and marketing contracts: a higher percentage of firms in the highest annual sales revenue category use these arrangements as compared to the other sales categories: (79% for the $5 million or more vs. 52% and 47% for the next two lower categories, respectively) and marketing contracts (70% for the $5 million or more vs. 55% [mid] and 45% [small]). A larger percentage of suppliers in the category of middle-sized sales revenue ($500,000 to $4,999,999) said that they use production contracts (64%) than those in the largest sales category (59%) or smallest sales category (47%) sales. Thus, it appears that suppliers with middle-sized sales revenues were more likely to engage in risk-sharing contracting arrangements than large or small suppliers were. In general, the largest respondents are more likely to utilize multiple contracting arrangements. Twenty-eight percent of large suppliers used all four, while only 14 percent of mid-size firms and 8 percent of small firms did.

The survey asked suppliers how important different contract characteristics are to them (Figure7-11a). A prior positive experience with a buyer and delivery timeline were both identified as "important" or "somewhat important" to over 80 percent of suppliers. A price guarantee and the guaranteed purchase of a predetermined quantity of seeds or plant material were similarly important. In contrast, a much smaller proportion of suppliers (35% and 29%, respectively) found that multi-production-cycle contracts or production cost-sharing was important or somewhat important. Guaranteeing a sufficiently high price can substitute for sharing the expected production cost. However, a higher price may not necessarily substitute for sharing production risk, which depends on the relative magnitude of price compared to production cost and also the supplier's attitude toward risk. Due to the substitutability across price guarantee, purchase quantity guarantee, and production cost-sharing that are charac-teristics of marketing and/or production contracts, a substantial majority of respondents deemed market factors a significant consideration.

Responses to this question were also disaggregated by the location and sales revenue categories of the supplier. Suppliers from the East and West were generally similar in their views on the importance of these characteristics. A slightly higher proportion of western suppliers (83%) than eastern growers (76%) indicated that a purchase guarantee for a predetermined quantity was important or somewhat important. As shown in Figure 7-11b, for most contract characteristics, suppliers in the highest sales category were most likely to state that the characteristic was important or somewhat important. However, approximately three out of four or more respondents in all sales revenue categories said that price guarantee, purchase guaranteed for predetermined quantity, delivery timeline, and prior positive expertise with the buyer were somewhat or very important characteristics.

When suppliers were asked about the timing of signing a contract relative to when they would start to produce native seeds or plant material for sale, about half (47%) of all suppliers indicated they normally sign a contract after seed production is complete. This suggests they have seed stored and ready for testing and delivery.

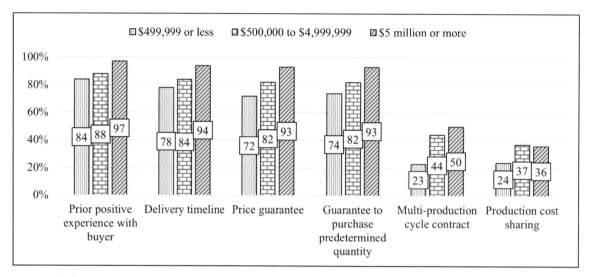

FIGURE 7-11b Importance of contract characteristics to suppliers, by annual sales (very important or somewhat important responses).

Figure 7-12 shows some nuances for groups of suppliers for the three categories of overall sales, and this might further vary by plant species. The first two sets of bars in Figure 7-12 should be interpreted with caution, as suppliers were asked about "foundation seed" as a proxy for any seed used as starting material for production; however, the term has specific meaning with respect to seed certification.

Business Activities of Suppliers—Anticipating Demand

Suppliers were asked about the role different types of information play in how they anticipate future demand for native seed or plants in order to plan ahead. Figure 7-13 shows that "past purchases or requests from buyers" plays a major (56%) or moderate (28%) role for nearly all suppliers. More than half said that "demand for urban

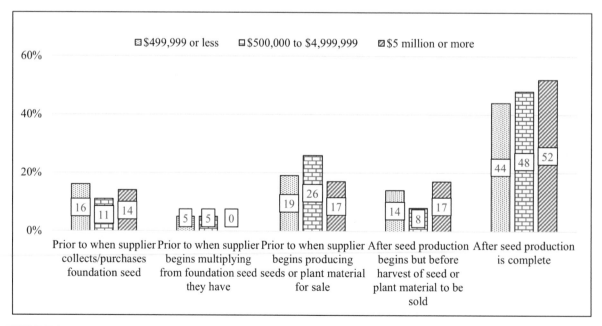

FIGURE 7-12 Typical timing of when a contract is signed, by annual sales.

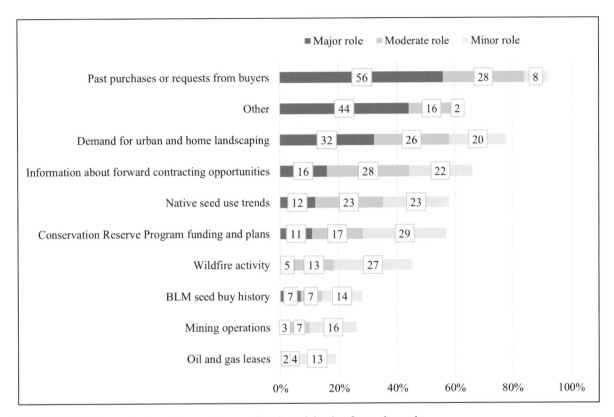

FIGURE 7-13 Types of information used by suppliers in anticipating future demand.

and home landscaping" plays a major (32%) or moderate (26%) role. The broad range of responses in the "Other" category reflected suppliers' interest in monitoring developments and opportunities related to native seed and plants both locally and nationally. Notably, the Conservation Reserve Program administered by the US Department of Agriculture (USDA) plays major and moderate roles for a larger share of suppliers in our survey than either wildfire activity or Bureau of Land Management (BLM) seed buy history.

Not being able to accurately anticipate demand can result in unsold seed, and suppliers were asked to provide an estimate of approximately what percentage of their inventory of native seed and plants was left unsold at the end of the marketing year during 2017–2019. Seventy-seven percent of the responding suppliers said that the percentage of native seed and plants left unsold was about what they anticipated. About 10 percent indicated that there was more left than anticipated, and 13 percent indicated that it was less than they had anticipated. Figure 7-14 shows the approximate percentage of inventory unsold at the end of the marketing year.

Interestingly, pairing these responses with buyers' responses about the availability and price of desired native seed and plants suggests that suppliers are able to market seed by anticipating buyers' willingness to substitute. Moreover, demand for seed that is an acceptable second choice for many buyers may be more predictable than demand for more specialized seed, which is more limited and potentially more variable. Of the respondents who replied that less than or about what they anticipated was left unsold, 44 percent considered insufficient demand a major or moderate challenge and 74 percent deemed unpredictable demand a major or moderate challenge. Another potential reason that the great majority of suppliers were on target at selling inventory may be that most have adequate storage (see below) and planned for inventory carryover.

Responses were examined by the three categories of overall sales. Among middle-sized suppliers ($500,000 to $4,999,999 in sales) 18 percent indicated that there was less left unsold than they anticipated, compared to 10 percent among small-size suppliers and 14 percent among large suppliers. It is not immediately clear why middle-size suppliers had less unsold than they anticipated. This could be related to the markets served and business models of middle-sized companies.

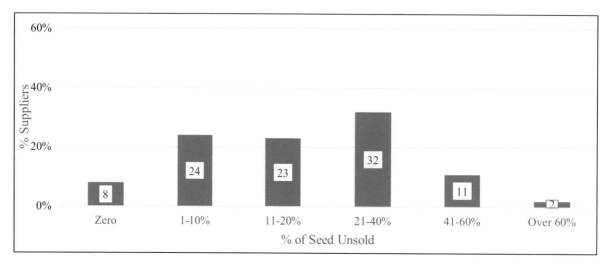

FIGURE 7-14 Approximate percentage of inventory unsold at the end of the marketing year.

Suppliers were asked about the extent to which the availability (or lack thereof) of different types of storage (freezer storage, refrigerated storage, and ambient storage) limits the quantity of seed they can sell. Approximately 30 percent of the responding suppliers said that the lack of freezer storage and/or refrigerated storage (32%) limits their sales (Figure 7-15a). There were no differences between how suppliers in the East and West answered this question. However, as shown in Figure 7-15b, storage is more of a limiting factor for quantity of seed potentially sold relative to the size of the suppliers (based on annual sales). In particular, nearly 40 percent of smaller firms said that they are constrained by limits on refrigerated storage.

CHALLENGES FACED BY SUPPLIERS

Respondents were asked about the challenge they face as suppliers of native seed or plants. Figure 7-16a shows that unpredictable demand was described as a major or moderate challenge by 74 percent of suppliers. Given the large share of suppliers who responded that the quantity of native seed or plant materials left unmarketed

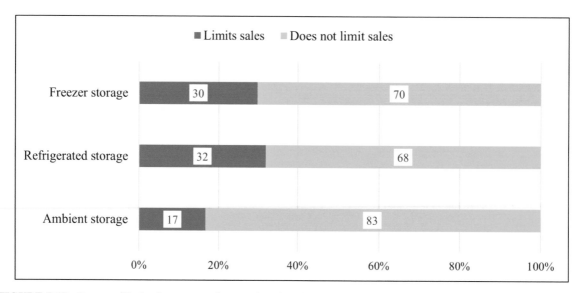

FIGURE 7-15a Impact of lack of storage on the quantity of seed suppliers can sell, by storage type.

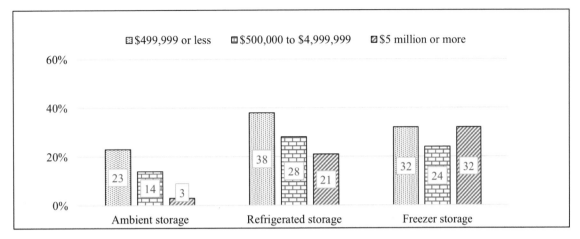

FIGURE 7-15b Impact of lack of storage on the quantity of seed suppliers can sell, by storage type and annual sales.

was either about or less than what they expected, this suggests that they believed that their sales could have been higher if they had been better able to anticipate demand for particular species or species by seed zone. Over half of the respondents (60%) indicated that "difficult-to-grow species" are a major or moderate challenge, and half (49%) mentioned the lack of stock seed from appropriate seed zones or other specified locations are a major or moderate challenge. Insufficient demand was a major or moderate challenge for a little under half of the suppliers (46%). As with responses regarding unpredictable demand, the responses about insufficient demand are likely based on specific species or species by seed zone rather than overall seed output. Fewer respondents mentioned lack of seed testing protocols (34%) and demand for plants without propagation protocols (24%) as a major or moderate challenge.

This question was also summarized by total sales categories and location of supplier, Figures 7-16b and 7-16c, respectively. Generally, larger suppliers and suppliers in the West were more likely to describe most of these

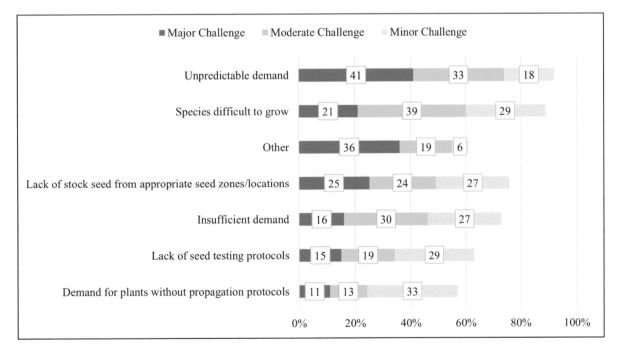

FIGURE 7-16a Challenges faced by suppliers of native seed and plant materials.

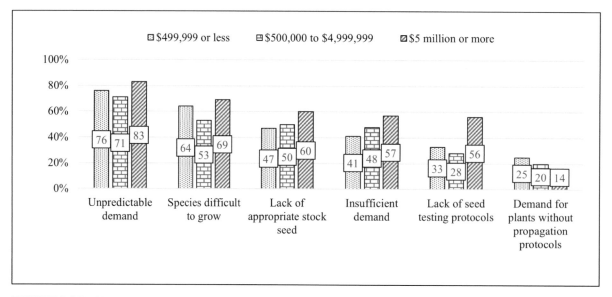

FIGURE 7-16b Challenges (major or moderate) faced by suppliers of native seed and plant materials, by annual sales.

challenges as a major or moderate problem. Demand for plants without propagation protocols seemed to be an exception that was more likely to be a challenge to smaller suppliers and suppliers in the East.

Ability to Expand Capacity

To better understand capacity, suppliers were asked whether they could expand their operations in response to anticipated higher demand. Specifically, suppliers were asked about ability to (1) wild-collect more seed; (2) grow more native plants with the goal of producing and selling native seed; and (3) grow and sell more native plants. Suppliers were also asked about perceived barriers and disincentives that might stand in the way of expanding in these areas.

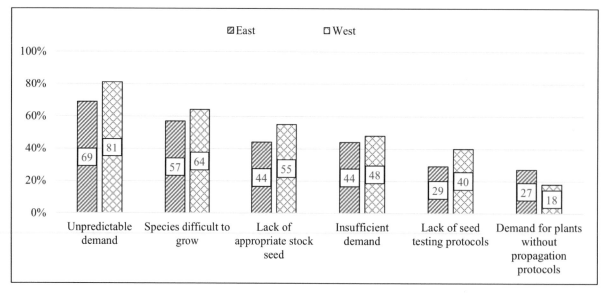

FIGURE 7-16c Challenges (major or moderate) faced by suppliers of native seed and plant materials, by region.

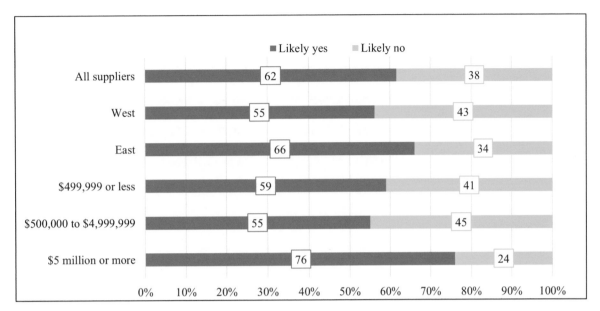

FIGURE 7-17 Suppliers' ability to expand wild collection of seed if they were to anticipate higher demand, by region and annual sales.

In response to the question "If you were to anticipate higher demand for directly wild-collected native seed, would your business be able to expand to collect more seed?" Sixty-two percent of suppliers said, "Likely yes" and 38 percent answered "Likely no." A higher percentage of suppliers in the East (66%) thought they could expand their business than in the West (55%). Seventy-six percent of the suppliers with the highest annual sales felt they could expand their wild-seed collection efforts. Figure 7-17 presents responses to this question for all suppliers, and by the overall sales categories and by the location of the supplier.

When asked about barriers or disincentives to wild-collecting native seed, 71 percent of suppliers said that major barriers exist. Table 7-1 provides a summary of the types of barriers that were described by the respondents and the number of respondents that mentioned this barrier. Because access to land is a barrier

TABLE 7-1 Supplier Perspectives on Major Barriers and Disincentives to Wild-Collecting Native Seed

Barrier	Count
Access to land	62
Availability of seed on the land	26
Labor	22
Depletion of the resource	17
Cost	13
Time	12
Markets	10
Knowledge	10
Environmental factors	7
Ethical reasons	6
Quality of seed	6
Resources available	4
Other	2

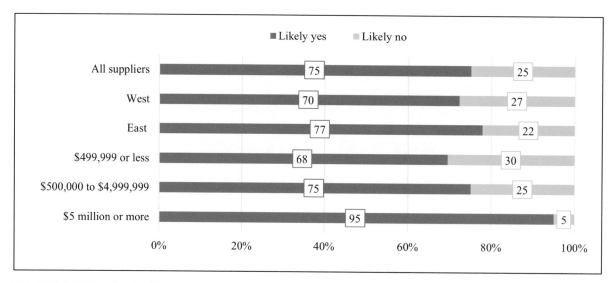

FIGURE 7-18 Suppliers' ability to grow more native plants with the goal of producing and selling native seed if they were to anticipate higher demand, by region and annual sales.

for many, the largest suppliers may have more access to private land from which to collect than companies with lower revenues.

In response to the question "If you were to anticipate higher demand, would your business be able to expand to grow more native plants with the goal of producing and selling native seed?", 75 percent said "Likely yes" and 25 percent said "Likely no." Larger suppliers were more likely to respond affirmatively. Figure 7-18 presents responses to this question for all suppliers, and by the overall sales categories and by the location of the supplier.

When asked about barriers to growing native plants with the goal of producing and selling native seed, 43 percent of suppliers said that there were major barriers. Table 7-2 provides a high-level summary of the types of barriers

TABLE 7-2 Supplier Perspectives on Major Barriers and Disincentives to Growing Native Plants with the Goal of Producing and Selling Native Seed

Barrier	Count
Markets	24
Seed/plant viability	14
Cost	14
Availability	8
Space	8
Environmental factors	7
Labor	6
Knowledge	6
Time	5
Resources	4
Zoning Requirements	3
Permits	3
Other	3

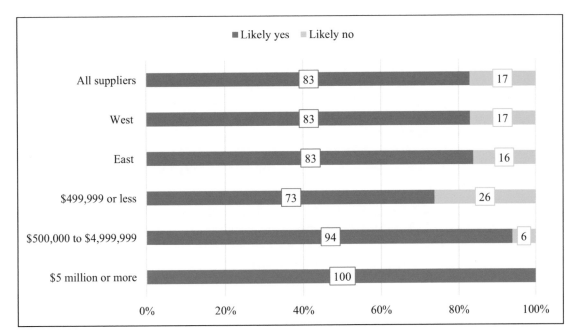

FIGURE 7-19 Suppliers' ability to grow and sell more plants if they were to anticipate higher demand, by region and annual sales.

that were described by the respondents and the number of respondents that mentioned this barrier. Twenty-four cited reasons categorized as "market," 14 cited "cost," and 8 cited "availability." Respondents may have cited more than one barrier. As noted earlier, cost and availability considerations are jointly driven by market forces. So, the dominant category of barrier to greater native seed supply appears to be the market-driven confluence of uncertain demand, high production costs, and availability of native plants as an input to the production process.

In response to the question "If you were to anticipate higher demand, would your business be able to expand to grow and sell more plant materials?", 83 percent of suppliers said "Likely yes" and 17 percent said "Likely no." The responses were similar among suppliers in the eastern and western states. The smallest suppliers were

TABLE 7-3 Supplier Perspectives on Major Barriers and Disincentives to Growing and Selling Plants

Barrier	Count
Markets	29
Knowledge, education, perception	13
Cost	10
Viability	9
Labor	8
Availability	8
Time	3
Space	3
Environment	4
Permits	1
See previous comment	5
Other	6

somewhat less likely to say that they could likely expand. Figure 7-19 presents responses to this question for all suppliers, and by the overall sales categories and by the location of the supplier.

In terms of barriers, 45 percent of the suppliers said that there were major barriers or disincentives to growing and selling plants. Table 7-3 provides a summary of the types of barriers that were described by the respondents and the number of respondents that mentioned this barrier. Similar to the types of barriers that surfaced in the responses about growing native plants with the goal of producing and selling native seed, the characteristics of the market were seen as the top barrier to growing and selling plants as well. Twenty-nine cited reasons categorized as "markets," 10 cited "cost," and 8 cited "availability."

SUGGESTIONS FOR ADDRESSING THE CHALLENGES

The survey asked suppliers to provide suggestions that might help address the challenges they have encountered as suppliers of native seeds or plant materials. Table 7-4 provides a high-level summary of the types of suggestions and the number of respondents that provided these comments.

DISCUSSION

Suppliers are pivotal to the native seed supply chain. The strongest message that came from the survey was that suppliers need greater consistency and predictability from their buyers. In an open-ended question asking for final thoughts, some specific suggestions were that buyers should plan projects and communicate their seed needs on realistic timelines that account for the time needed to acquire and/or propagate material; that an information clearinghouse could be a useful mechanism for better connecting buyers and suppliers, thus reducing market friction; and that risk-sharing mechanisms such as contracting or down payments could help create a more stable business environment for suppliers. Several respondents pointed out that an unstable market is especially disadvantageous to small growers, who are important suppliers of diverse locally adapted native species.

TABLE 7-4 Supplier Suggestions to Address the Challenges Encountered

Category	Description of the Types of Suggestions in the Category	Count
Planning	Better planning about native seed or plant needs (includes comments about how long it takes to grow a plant before it is ready to sell and disconnect between planning needs and time for a plant to grow).	31
Awareness, education	Need for more and/or better education around native seeds or plant materials.	25
Coordination	Better coordination or communication between entities.	20
Source	Need for more native seed/plant sources, and issues related to availability and production.	15
Funding	Need for more funding and/or consistent funding.	14
Policy	Changing policies that impact growing seed.	10
Markets or economics	Addressing economic or market concerns	10
Value shift	Need to change paradigms, philosophy, mindset, etc.	9
Seed specific	Suggestions specific to seed/climate change/germination/growing situation (coded as seed and not planning when it is about a specific type of seed or plant or crop).	8
Regional focus	More focus on regional/local seed production and/or sharing of collections.	7
Standards	Need for better standards, uniformity.	7
Labor	Addressing labor needs or requirements.	5
Access	Addressing issues of access (without reference to policies).	4
Other	Other responses not categorized elsewhere.	2

The suppliers who responded to this survey most often anticipated demand not trying to predict wildfire frequency or what was previously requested by the BLM consolidated buys, but by more consistently growing markets such as the urban and landscaping sector, and the USDA Conservation Reserve Program. That does not mean that the seeds they supply to these markets would be appropriate for all restoration purposes, only that markets that are predictable and stable are able to get the focus of suppliers.

Another theme of the open-ended comments was that buyers are not always knowledgeable about the use of native plants in restoration and other uses. They may not be aware of which native plants to use in a given project, how to propagate these materials successfully, or even why it is important to use native species in restoration. There is confusion around seed zones, with some buyers being unaware of the need to use genetically appropriate material, while others have unrealistic expectations about how local the material should be. Better education of the personnel involved in seed buying and restoration, greater collaboration between buyers and knowledgeable suppliers, and a clear and agreed-upon system of seed zones were among the suggestions in this area. Even more broadly, suppliers suggested, a more thriving seed industry could emerge if agencies and their contractors, and the general public, had a better understanding of the importance of native plant species.

Multiple suppliers commented on the need for improved access to native seed collection sources on public lands, calling for a more streamlined permit process, protection of important source areas, and even the potential use of successfully restored sites for seed collection. Several growers also called for Plant Materials Centers to play a larger role in developing and distributing stock seed. The lack of stock seed of the local ecotypes desired by buyers is a barrier to their supply. Given that about two-thirds of suppliers sell seed of claimed wild origin, getting certified seed stock seed into the hands of producers would increase the confidence of seed users that the origins of the seed purchased has been verified.

In addition to establishing clear seed zones and an information clearinghouse, another function that suppliers suggested could be an improved system of seed testing, quality control, and labeling, which would protect high-quality suppliers. Some suppliers also called for technical assistance, such as the development of techniques for cleaning and propagation.

CONCLUSIONS

Conclusion 7-1: *Suppliers view unpredictable demand as their leading challenge (75% of respondents). The problem is particularly acute among western suppliers of native seeds and plants (81%).*

Conclusion 7-2: *The top two technical challenges reported by suppliers are difficult-to-grow species of native plants (60%) and lack of stock seed from appropriate seed zones or locations (50%).*

Conclusion 7-3: *If they could reliably anticipate demand, the majority of suppliers who replied to these questions said that they could (1) expand wild collection of native seed (62%), (2) produce more native seed (75%), and (3) grow more native plants (83%).*

Conclusion 7-4: *Suppliers use past purchases to predict future demand, although suppliers also know that buyers will take substitutions, which creates a circular system that does not serve to expand the diversity of species and ecotypes for the market.*

Conclusion 7-5: *For increasing the supply of wild-collected native seed, the principal barrier is access to land. For expanding production of native seed and plant materials, 43 to 45 percent of suppliers reported barriers, with markets cited as the leading barrier in both cases.*

Conclusion 7-6: *Sixty-five percent of suppliers report that they sell certified Source-Identified seed. About the same percentage (68%) also sell seed of claimed wild origin or provenance.*

Conclusion 7-7: *Lack of refrigerator or freezer storage limited sales for roughly 30 percent of native seed suppliers.*

Conclusion 7-8: *Among supplier suggestions to address the challenges that they encounter, four of the top seven response categories were related to communicating demand to suppliers (i.e., planning, communication, funding, and markets or economics).*

Conclusion 7-9: *Some suppliers suggested that a more thriving seed industry could emerge if agencies and their contractors, and the general public, had a better understanding of the importance of native plant species.*

Conclusion 7-10: *Marketing or production contracts were used by half of supplier respondents. The most highly valued contract characteristics (described as "very important" or "somewhat important" by 75% or more of suppliers) were delivery timeline, price guarantee, and guarantee to purchase a predetermined quantity.*

8

Knowledge Gaps and Research Needs to Support the Native Seed Supply

For the seed supply to work efficiently, there is a need for a strong foundation of knowledge to underpin decision making. Because native seed is a high-value product and decisions about its use have consequences, not the least of which is the success or failure of restoration projects, the effective management of the nation's seed supply must be supported by knowledge drawn from many sources.

Over the course of the assessment the committee learned of multiple types of information needs. There are some issues that call for innovative and even transformative research with potentially broad applicability across the nation, for example, seed sourcing with particular attention to climate change, the integration of traditional ecological knowledge into restoration practices, and the economics of native seed production and markets.

Other issues call for the accelerated development and wider dissemination of improved technical knowledge relevant to specific regions and/or species, including how to improve the genetic diversity, viability, quality testing, long-term storage, and deployment of seeds and other native plant materials for restoration. While not a research topic per se, the utility of seed certification is highlighted (Box 8-2) as an avenue whereby the genetic research suggested in this chapter can be supported with a consistent set of definitions of germplasm status and development and cultivation protocols.

SEED SOURCING AND SEED ZONE DELINEATION

There is a considerable body of research showing that plant populations can become locally adapted to climate, soils, competitive regimes, pests and pathogens, and many other factors (Hufford and Mazer, 2003; Linhart and Grant, 1996). This has led to local seed sourcing preferences for restoration, although the definition of "local" can be quite divergent between projects. For several species, seed transfer zones have been empirically derived from common garden studies. A summary of these, with maps, data, and citations, can be found on the US Forest Service's Threat and Resource Mapping site.[1] Seed transfer zones are geographic areas within which seeds can be planted with minimal risk of maladaptation (Kramer and Havens, 2009). Local adaptation can vary significantly between species so seed transfer zones are ideally determined on a species-by-species basis (Johnson et al., 2010; Leimu and Fischer, 2008).

For taxa where seed transfer zones have not been studied, provisional seed zones have been developed (Bower et al., 2014; Doherty et al., 2017) (see Figure 8-1a) including those that augment climate-only zones with molecular genetics (Massatti et al., 2020) and arbitrary geographic distance (Saari and Glisson, 2012) to source seeds.

---

[1] See https://www.fs.usda.gov/wwetac/threat-map/TRMSeedZoneMapper.php (accessed February 1, 2023).

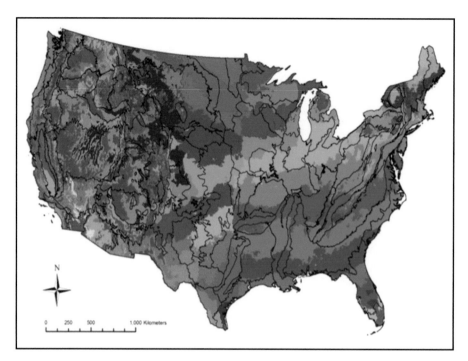

FIGURE 8-1a Provisional seed transfer zones (colored areas) developed by the US Forest Service for the continental United States with an overlay of Omernik Level III ecoregional boundaries (black outlines) that distinguish areas with similar climate, but that differ ecologically.
SOURCE: US Forest Service (Bower et al., 2014; Omernik, 1987).

However, many studies suggest that using an ecologically similar source site leads to better outcomes (Hereford, 2009; Johnson et al., 2010, 2015). See Figure 8-1b for a map of Omernik Level III ecoregions used to source seeds for all species.

Despite the existence of considerable case-by-case research on local adaptation, there is still a great need to improve the overall scientific basis for seed zone definitions, with the goal of creating widely accepted standards for seed zone delineation which could significantly improve the functioning of the native seed market.

SEED SOURCING AND CLIMATE CHANGE

Rapid climate change now poses a fundamental challenge to traditional approaches to seed sourcing in restoration. Many restoration projects are now obtaining seed from more distant sources (provenances), employing strategies such as broadening the definition of local (relaxed local provenancing), creating mixtures from different source areas (composite provenancing, admixture provenancing), and obtaining seeds from regions where the climate resembles the predicted future climate of the restoration location (predictive provenancing) (Breed et al., 2013, 2018; Broadhurst et al., 2008; and Havens et al., 2015).

Mixed-provenance strategies have also been proposed to increase genetic diversity and therefore the potential for climate adaptation (Bucharova et al., 2019; Prober et al., 2015; but see Kramer et al., 2018). The Climate Smart Restoration Tool[2] maps current and future seed transfer limits for seed lots based on empirical or provisional data.

The science of climate-adapted restoration is new and rapidly evolving, and conclusive empirical tests of proposed strategies are scarce. This creates tremendous needs and opportunities for basic and applied research, which could include conducting an actual restoration project in a rigorous experimental fashion using the principles of adaptive management. We note that seed of known provenance (such as source-identified seed) is a critical requirement for such research (Havens et al., 2015; Vitt et al., 2022).

---

[2] See https://climaterestorationtool.org/csrt/ (accessed December 1, 2022).

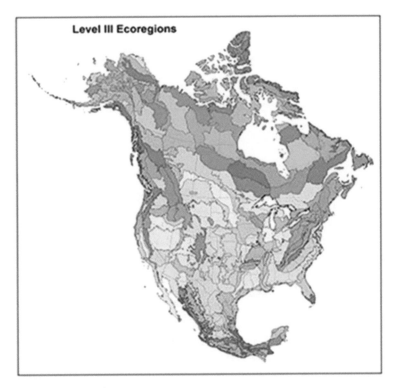

FIGURE 8-1b Omernik Level III ecoregions.
SOURCE: Environmental Protection Agency (Omernik, 1987).

SPECIES DIVERSITY AND COMPOSITION

Just as seed selection will be informed by studies on the role of genetic diversity and provenance within a species and population, there are many research questions to be addressed about the appropriate selection of species and species mixes for restoration projects. Of course, the basic goal in many cases is to reestablish as many as possible of the plant species that were present at a site prior to disturbance. However, there are other instances in which this goal is not achievable because of highly altered site conditions, lack of pre-disturbance information, or scarcity of the necessary plant material. Also, there are cases in which the selection of species mixes is directed at outcomes such as supporting pollinators or wildlife, regenerating depleted soil microbiota, or maintaining stable plant cover under challenging physical conditions. For selecting species mixes, an extensive literature based primarily on controlled small-scale experiments suggests the general principle that higher levels of plant diversity at the species, functional, and/or phylogenetic levels will lead to higher average biomass, temporal stability of biomass, or other desirable aspects of ecosystem function (Cardinale et al., 2006, 2007, 2012; Gamfelt et al., 2013; Gross et al., 2014; Hector and Bagchi, 2007; Hooper et al., 2012; Kinzig et al., 2002; Loreau and Hector, 2001). The application of this ecological principle to actual restoration practice, especially at the scales of heterogeneous landscapes, is highly deserving of further research. Many other questions concerning species composition in relation to restoration success are also in need of further attention. For example, either functional traits (e.g., Leger et al., 2021), or patterns of species co-occurrence in undisturbed sites (e.g., Agneray et al., 2022), may be valuable tools for designing species mixes that are capable of achieving such goals as withstanding harsh environments, co-occurring stably, and resisting invasion.

Research on general principles of restoration success is the provenance of the federal research agencies, including the National Science Foundation, US Department of Agriculture National Institute of Food and Agriculture (NIFA), and the US Geological Survey (USGS). For example, applied research being pursued by the USGS in the context of the Colorado Plateau Plant Program seeks answers to fundamental questions with practical applications (see Box 8-1).

BOX 8-1
Research Agenda for the Colorado Plateau Native Plant Program

The US Geological Survey (USGS) is conducting applied research for the Colorado Plateau Native Plant Program, a collaborative partnership with the Bureau of Land Management (BLM), and long list of partners in academe, industry, and nongovernmental organizations. Using restoration activities as a platform for experimentation, the Survey's work in the arid Plateau will provide insights that are important for seed selection and restoration success. For example, the USGS research agenda for 2021–2022 has set out the following research tasks, seeking answers to practical questions:

- Resolve patterns and drivers of genetic diversity, structure, and adaptation (i.e., landscape genetics): *How does the degree of genetic diversity (too diverse, too little diversity) in plant materials used for restoration affect restoration outcomes?*
- Determine adaptive phenotypic variation in natural populations (i.e., common gardens and plant traits): *Can experimental plantings of a species sourced from different locations in a prospective new site provide clues to local adaptation?*
- Quantify seed survival and establishment in the context of growing aridity: *If soil moisture is critical for seed establishment, how will weather variability under climate change affect regeneration success in restoration projects?*
- Investigate the impact of seed increase on the genetic identity of restoration materials: *How does the size of an accession used for seed increase affect the genetic identity of seeds relative of the wildland population?*
- Investigate the long-term impacts of restoration materials on the genetic identity of plants in their natural communities: *How do "non-local" plant materials affect pre-existing plant communities?*

TRADITIONAL ECOLOGICAL KNOWLEDGE

Indigenous cultures traditionally place tremendous importance on the links between human and nonhuman domains, which is reflected in the diverse practices with which they historically managed their plant and animal resources (e.g., Hornborg, 2006; Viveiros de Castro, 2004). Former federal policies reducing tribal land and forced cultural assimilation threatened traditional knowledge for much of the 19th and 20th centuries (see Chapter 5). Recovering and maintaining traditional knowledge was of high importance in the Tribal Nursery Needs Assessment (Luna et al., 2003).

For land management and native plant restoration, there is a growing realization that combining traditional ecological knowledge with western science can benefit both tribal needs and western science approaches to native plant research (Anderson and Barbour, 2003; Berkes, 2008; Berkes et al., 2000; Dockry et al., 2022; Eisenberg et al., 2019; Pierotti and Wildcat, 2000), a theme also highlighted in recent popular books (e.g., Kimmerer, 2013). The Fort Belknap Native Seed and Restoration Program and the Grand Ronde Tribal Native Plant Materials Program described in Chapter 5 show how native knowledge and western science can successfully engage.

A critical area for research is to better understand the collecting, growing, harvesting, and storage of native plants on tribal lands, in the context of managing and promoting ecological restoration and ecosystem management and the cultural uses of native plants. This need was also noted in the Tribal Nursery Needs Assessment and the Nursery Manual for Native Plants (Dumroese et al., 2009), yet it remains largely unmet. Critical to this process is engaging tribes in all aspects of planning, carrying out, and applying results from scientific projects to promote ecosystem management and native plant restoration (Dockry et al., 2022; Farley et al., 2015).

ECONOMICS OF NATIVE SEED PRODUCTION AND MARKETS

This report offers a general sense of availability and uses of native seed and plant materials, but data gaps prevented it from offering a clear, quantitative picture at the national level. Not only is the magnitude of total native seed use currently not clear, by extension so too is our overall quantitative understanding of how native seeds are used and the costs to produce them. Economic research is needed to understand (1) the nature of demand and supply in the current native seed market; (2) the evolution of native seed uses over time; (3) costs of production for major types of native seed; and (4) sources of production and price risk, as well as the effectiveness of alternative contract designs at mitigating those risks.

A comprehensive market study is needed simply to characterize the magnitude and details of the US native seed and plant material market. As noted in the introduction to this report, one-third of US territory is in government hands, with the rest held privately (Figure 1-1). For certain federal agencies, land areas under direct management are well known (Table 3-1), and some federal, state, and tribal agencies can report the use of native seed on their lands. Although tracking native seed use on private agricultural land is trickier because of the many buyers, the area planted with native seed under USDA programs like the large Conservation Reserve Program (see Iovanna and Pratt presentation reported in Chapter 6) is fairly clear (although seed species and quantities are not). On non-agricultural private land, there is no central data source to estimate native seed use. So, the first step is to measure the total land area planted with native species and the annual demand for native seed and plant materials.

As this report has shown, the uses of native seeds and plant materials vary widely. Some users aim to restore native plant ecology, while others seek to serve specific functions, such as soil conservation, pollination, or aesthetics. In effect the larger market for native seeds is composed of many submarkets of seed users with specific preferences. Users in some of those submarkets are more willing to substitute one species for another, meaning that they are more price sensitive. Research in the Colorado Plateau (Camhi et al., 2019) has found that where users are more open to substitutions (e.g., to meet a particular ecosystem function, like supporting rangeland grazing), they are more likely to choose seed that meets that goal at a lower price. By contrast, where seed demand must meet rigid specifications (e.g., to restore plant ecology with locally adapted genotypes), buyers are less willing to change species because of prices (so demand is more "price inelastic" in economist parlance). A quantitative, national assessment of native seed user requirements is needed to characterize those submarkets and how best to meet the component demands. Such a study should be done on a regional basis, not only because of the difference in land ownership regionally (e.g., more public land in the West), but also because certain uses are more urgent than others (e.g., establishing native rangeland species ahead of invasives versus planting roadsides with pollinator-friendly species).

Understanding future demand for native seeds calls for understanding past patterns. As wildfires have covered more acreage in the United States, the demand for seed to reestablish vegetative cover has grown. In recent years, it has become higher priority to meet that demand with native species—rather than exotics. On agricultural land too, the emphasis on native species has certainly expanded under the USDA Conservation Reserve Program. Thoughtful projection of future native seed needs will require an understanding of past patterns, both quantitatively and by species. This understanding must be paired with an awareness of how climate change is altering the demands for native seed.

Complementing demand-side studies there is a need to understand better the cost and time needed to produce native seed and plant material. Such information underpins the economic supply that makes native seed available. Here again, there is a need to study market segments, as costs of production will be strongly affected by such factors as the need for locally adapted genetics, availability of wild seed collection sites, plant life cycle (perennial versus annual), seed yields, and scale of production. Such production cost research would provide the basis for survey research on producer willingness to supply native seed under varying conditions.

Risk management is a high priority for producers of native seed and plant materials, as reported in the supplier survey here. Much risk research is needed at the regional level for key native species. Research to improve understanding of production and price risks would inform better designs for production and marketing contracts. To date, there have been important innovations, such as the BLM Indefinite Delivery Indefinite Quantity contracts and the USFS Blanket Purchase Agreements, but there exists a plethora of alternative contract designs that could be adapted to meet the needs of the market for native seeds. Separate from characterizing native seed production and market risks is the need for research to test the alternative contract designs at connecting market demand with producer supply.

BASIC SEED PRODUCTION INFORMATION FOR MORE SPECIES

Species-specific applied research is needed to improve the protocols for planting, growing, and harvesting many species. For many native grasses, production protocols are available and large quantities of seed can easily be produced using standard farming equipment. Native forbs are more problematic as they belong to a variety of plant families, with variable fruit and seed morphology and timing of ripening. Seed production may not begin until the second year or beyond, pollinator requirements may be unknown or may include native pollinators not present at the farm site, and insect and disease issues may emerge when the species is grown in a seed field monoculture (Cane, 2008; Shaw and Jensen, 2014). Very basic planting requirements (seeding date, rate, depth, etc.) must be determined, and specialized planting or harvesting equipment may be needed, when bringing new species into production. Stock seed supplies are often limited, necessitating an initial increase before field production can begin. The results of on-farm trial and error are usually unavailable to other growers of the same species. Conducting and disseminating applied research has considerable potential to expand the diversity and volume of native seeds for restoration.

MAINTAINING GENETIC INTEGRITY DURING CULTIVATION

Maintaining the genetic integrity of seed through the increase process is also an important concern. Native seed farming to increase amounts of seed sourced from wild populations aims to maintain genetic diversity but may cause unintended genetic changes, particularly when plants are grown under conditions quite different from the wild source population. Cultivated populations can become maladapted to conditions in the wild (Ensslin et al., 2015; Espeland et al., 2017; Havens et al., 2004; Husband and Campbell, 2004). In production, plants may receive irrigation, be protected from herbivores, and be grown in rich or fertilized soils and under novel climatic conditions that could make them less fit when reintroduced (Conrady et al., 2022; Espeland et al., 2017). For example, cultivated *Carlina vulgaris* and *Jasione montana* exhibited loss of tolerance for drought and competition (Ensslin et al., 2015), and farmed *Clarkia pulchella* (of non-certified seed) showed strong increases in mortality especially in simulated drought conditions (Pizza et al., 2021), although some other studies found less evidence for cultivation-induced losses of genetic integrity of native seed (Conrady et al., 2022).

Seed certification requirements (see Box 8-2) are designed to reduce genetic changes by limiting the number of generations that can be produced from stock seed. Since generations grown, species mating system, the production environment, and on-farm methodology all may differ and affect seed genetic integrity, more species-specific research is needed on genetic changes and how they can be managed during agronomic production.

SEED BIOLOGY AND SEED ANALYSIS

Accurate information about seed quality is essential for setting seed prices and calculating seeding rates. Seed tests that do not represent accurate seed quality affect costs and project success for users and income for producers. There are two difficulties in achieving quality information. First, there are federal, state, and private entities involved with analyzing and labeling native seed sold in the marketplace. Both regulations and laboratory quality across entities can be uneven. Second, the tests needed to assess quality are difficult to conduct accurately.

State seed laboratories in most states are members of the Association of Official Seed Analysts (AOSA)[3] and follow AOSA seed analysis rules. State seed acts and regulations provide for truth in labeling and enforcement for seed lots of species covered by AOSA rules sold within the state. These requirements may differ across states. The Federal Seed Act (FSA) (USDA 1940 with revisions)[4] takes precedence for seeds sold in interstate commerce but includes few native species. In addition to state seed laboratories there are commercial and private seed laboratories and a federal regulatory seed laboratory. There are several seed testing and accreditation programs to which seed laboratories and individual seed analysts may apply for accreditation.[5] This section discusses these issues and newer technology that may help with solutions.

---

[3] See https://analyzeseeds.com/about-us (accessed February 10, 2023).

[4] See https://www.ams.usda.gov/sites/default/files/media/Federal%20Seed%20Act.pdf (accessed February 10, 2023).

[5] See https://www.betterseed.org/resources/seed-testing-accreditation-schemes/ (accessed February 10, 2023).

Labeling seed for sale requires results of germination, purity, and noxious weed seed tests and that the date of each test be listed on the seed tag. Percent germination includes the germinated, hard, and dormant seed in the tested sample. Purity testing requires examination of 2,500 seeds to determine the percent pure seed, other crop seed, weed seed, and inert material. A sample of 25,000 seeds is required for noxious weed testing and the species and number of seeds of each noxious species found must be listed. Seed lots cannot be transported through or offered for sale in a state if the seed lot contains a species that is either prohibited (Prohibited Weed Seed), or present in excess of the number per pound (Restricted Weed Seed) as designated in the state seed law. Seed lots cannot be sold if they contain any seeds of species designated as a restricted noxious in the state where sold. Testing must be conducted within the period specified by the FSA (if applicable) for interstate shipment and by the state where the seed is sold.

There are more than 900 native species with official AOSA germination rules (about 5% of the US native flora). For some other species there are unofficial protocols that have come into use but have not received formal review and acceptance. These are provided in an AOSA database: Test Methods for Species without Rules.[6] For the many native species lacking any guidance, analysts often use procedures found in the literature or rules or protocols developed for members of the same genus or for species from similar habitats. The result is that for these species, different procedures may be used by different laboratories, often resulting in disparate results.

Seed quality data are essential for setting seed prices and calculating seeding rates. Because both initial germination and total viability (includes dormant seed) of native seed lots are often highly variable, seed price and seeding rates are generally calculated on a basis of pure live seed (PLS) per pound. PLS pounds in a seed lot = [% purity/100] × [% viability/100] × pounds of bulk seed. Seed tests that do not accurately represent actual seed quality affect costs and project success for users and income for producers. In addition, seeding rates calculated on such inaccurate tests will not represent the actual PLS seeded per area, leading to misinterpretations of seeding outcomes and monitoring data.

Discrepancies in test results can arise from several factors:

- The number of Registered Seed Technologists at seed laboratories may be insufficient to meet the needs of the native seed industry.
- Many species lack official testing rules or protocols; thus different laboratories may use different test procedures and obtain differing results.
- Analysts often have little or no experience with many native species that are only rarely submitted for testing.
- Germination tests for dormant seeds requiring cold or warm stratification may take weeks or months. A viability test using tetrazolium chloride (TZ) is often requested as a substitute for a germination test as it can be completed relatively rapidly. The TZ tests do not reflect germination ability, but rather whether the seed is alive or respiring, or not (total viability). The tests must be conducted by experienced analysts and can represent additional costs.
- Equipment availability and quality (e.g., germinators, balances, microscopes, seed herbaria) varies among laboratories.

Solutions to these problems will require seed biology research, leading to the development of additional AOSA rules. Increased laboratory accreditation and training of analysts in native seed testing could also contribute to more uniform results among laboratories. Improved funding to enable acquisition of high-quality equipment would reduce differences among analysts in their ability to identify seeds and evaluate test results. Blind referee tests can identify discrepancies in test results among laboratories and point to training needs. New technologies being developed for crop seed testing and variety development may lead to more consistent results and require far less time than current methods that require examination of individual seeds. X-ray imaging in combination with ad hoc image analysis software can be used to automatically calculate fill and seed weight. By calibrating the interpretation of x-ray images to germination results, it would be possible to determine viability without the need for destructive and time-consuming tests, such as germination and TZ. Moreover, computer vision and AI technology applied to seed images can aid in seed identification and recognition of defects. These techniques require extensive image galleries for individual species, and the greater morphological variability in wild populations compared to highly selected crop varieties create challenges.

---

[6] See https://analyzeseeds.com/test-methods-for-species-without-rules/ (accessed February 10, 2023).

BOX 8-2
Seed Certification and Genetic Integrity

Seed quality testing and a tag listing the testing results are required legally to sell seed. However, seed laws do not currently require the verification of the source location of seed and other attributes that aid in identifying the best seed for a particular project or seed zone. Some private and public organizations procure seed from trusted collectors and growers without such verification, but for most native plant materials offered for sale, independent, third-party verification can be provided by state seed certification agencies (SCAs). This is an optional, fee-based service paid by plant materials producers, but the nominal cost is passed on to buyers desiring verification of source and production protocols that maintain genetic integrity.

As the genetic composition of "locally" sourced native plant material used for restoration ostensibly provides an adaptive advantage and promotes ecological relationships needed for desirable ecosystem services, the SCAs introduced a process to verify seed collected and increased from wild plant populations through "natural track" certification in the early 1990s. The categories on the natural track are Source Identified (SI), Selected (S), Tested (T), and Variety/Cultivar (see "How AOSCA Tracks Wildland Sourced Seed"[7]).

All natural-track germplasm is genetically non-manipulated and may be advanced from SI (geographic origin known) to S, T, and Variety/Cultivar categories. The latter designations may be used to indicate that the intact population from which the seed was collected is recognized to have desired traits and ecological attributes (e.g., drought tolerance, wide geographic adaptation, etc.) applicable to specific revegetation and restoration needs of seed buyers.

Seed zones to promote the use of locally adapted genetic material for ecological restoration are being developed and utilized although additional research and experience are needed. Seed zone or other geographic designations for single or pooled accessions of native plant populations can be accommodated by SCAs if they are provided with appropriate documentation as to source locations and collection procedures used.

Population changes and inadvertent selection with successive seed generations can alter the properties of plant materials. Although inevitable, they can be mitigated by ensuring that the initial population size collected and subsequent seed increase generations are adequate, conducting seed increases in similar environments to those of the original seed source, monitoring seed increases as part of the certification process, and long-term maintenance of seed sample from the initial G_0 population and all seed increases (G_x) for future comparisons and use as stock seed as necessary. Given the need for third-party verification and tracking of germplasm source, seed increase generations, genetic distinctions, and cultivation protocols, certification is a valuable tool to protect the interests of participants in the native seed supply chain.

DISCUSSION

Restoration ecology has existed as a scientific discipline since the first half of the 20th century, and has made many important strides, yet is also noted for suffering a substantial disconnect between science and practice (Burbidge et al., 2011; Cabin et al., 2010; Dickens and Sudling, 2013). In this chapter we have identified what we believe are the most important knowledge gaps that impede the ability of the native seed supply to develop more fully and function effectively.

Some of the knowledge gaps lie in the realm of investigator-driven, conceptually innovative research, which is traditionally supported by competitive funding from research agencies including the National Science Foundation and USDA-NIFA. To maximally influence the nation's seed supply chain, such basic research will need to be carried out at large scales, and using environments and species, that reflect the real-world challenges and constraints of restoration practice.

---

[7] How AOSCA Tracks Wildland Sourced Seed, see https://www.aosca.org/programs and services (accessed December 30, 2022).

Other critical knowledge gaps lie in technical areas that are typically addressed by applied research organizations including USGS, USDA Agricultural Research Service, and others. While much applied research has already been carried out on many of the subjects the committee has discussed, there remains a need to identify solutions that are broader and more generally applicable—a nationwide system of seed zones being a notable example.

Conclusion 8-1: *Many information gaps affect the ability of the native seed supply to function efficiently and effectively. Addressing them would inform decision making, reduce uncertainty, and improve restoration outcomes.*

Conclusion 8-2: *Important gaps in basic research include developing and testing strategies for seed sourcing that take rapid climate change into account; integration of traditional ecological knowledge into ecological restoration; and the role of species diversity, traits that contribute to survival, and the economics of the native seed industry.*

Conclusion 8-3: *Critical needs for the development and dissemination of improved technical knowledge include identifying basic growing requirements for more species, maintaining genetic integrity during cultivation, improved approaches to seed analysis, and greater use of seed certification.*

Conclusion 8-4: *Tribal leadership and collaboration are essential features of research on traditional ecological knowledge and tribal uses of native plant materials.*

REFERENCES

Agneray, A., F. Matthew, T. Parchman, and E. Leger. 2022. Does a history of co-occurrence predict plant performance, community productivity, or invasion resistance? BioRxiv. https://doi.org/10.1101/2022.09.20.508783.

Anderson, M.K., and M.G. Barbour. 2003. Simulated indigenous management: A new model for ecological restoration in national parks. Ecological Restoration 21(4):269–277.

Berkes, F. 2008. Sacred Ecology: Traditional Ecological Knowledge and Resource Management, 2nd edition. Routledge.

Berkes, F., J. Colding, and C. Folke. 2000. Rediscovery of discovery of traditional ecological knowledge as adaptive management. Ecological Applications 10(5):1251–1262.

Bower, A.D., J.B.S. Clair, and V. Erickson. 2014. Generalized provisional seed zones for native plants. Ecological Applications 24(5):913–919.

Breed, M., M. Stead, K. Ottewell, M. Gardner, and A. Lowe. 2013. Which provenance and where? Seed sourcing strategies for revegetation in a changing environment. Conservation Genetics 14:1–10.

Breed, M.F., P.A. Harrison, A. Bischoff, P. Durruty, N.J.C. Gellie, E.K. Gonzales, K. Havens, M. Karmann, F.F. Kilkenny, S.L. Krauss, A.J. Lowe, P. Marques, P.G. Nevill, P.L. Vitt, and A. Bucharova. 2018. Priority actions to improve provenance decision-making. BioScience 68(7):510–516.

Broadhurst, L.M., A. Lowe, D.J. Coates, S.A. Cunningham, M. McDonald, P.A. Vesk, and C. Yates. 2008. Seed supply for broadscale restoration: Maximizing evolutionary potential. Evolutionary Applications 1(4):587–597.

Bucharova, A., O. Bossdorf, N. Hölzel, J. Kollmann, R. Prasse, and W. Durka. 2019. Mix and match: Regional admixture provenancing strikes a balance among different seed-sourcing strategies for ecological restoration. Conservation Genetics 20:7–17.

Burbidge, A., M. Maron, M. Clarke, J. Baker, D. Oliver, and G. Ford. 2011. Linking science and practice in ecological research and management: How can we do it better? Ecological Management & Restoration 12:54–60. https://doi.org/10.1111/j.1442-8903.2011.00569.x.

Cabin, R.J, A. Clewell, M. Ingram, T. McDonald, and V. Temperton. 2010. Bridging restoration science and practice: Results and analysis of a survey from the 2009 Society for Ecological Restoration international meeting. Restoration Ecology 18(6):783–788. https://doi.org/10.1111/j.1526-100X.2010.00743.x.

Camhi, A.L., C. Perrings, B. Butterfield, and T. Wood. 2019. Market-based opportunities for expanding native seed resources for restoration: A case study on the Colorado Plateau. Journal of Environmental Management 252:109644.

Cane, J.H. 2008. Pollinating bees crucial to farming wildflower seed for US habitat restoration. In R.R. James and T.L. Pitts-Singer, eds. Bee Pollination in Agricultural Systems, pp. 48–63. Oxford, UK: Oxford University Press.

Cardinale, B.J., D.S. Srivastava, J.E. Duffy, J.P. Wright, A.L. Downing, M. Sankaran, and C. Jouseau. 2006. Effects of biodiversity on the functioning of trophic groups and ecosystems. Nature 443(7114):989–992.

Cardinale, B.J., J.P. Wright, M.W. Cadotte, I.T. Carroll, A. Hector, D.S. Srivastava, M. Loreau, and J.J. Weis. 2007. Impacts of plant diversity on biomass production increase through time because of species complementarity. Proceedings of the National Academy of Sciences of the United States of America 104(46):18123–18128. doi: 10.1073/pnas.0709069104.

Cardinale, B.J., J.E. Duffy, A. Gonzalez, D.U. Hooper, C. Perrings, P. Venail, A. Narwani, G.M. Mace, D. Tilman, D.A. Wardle, A.P. Kinzig, G.C. Daily, M. Loreau, J.B. Grace, A. Larigauderie, D.S. Srivastava, and S. Naeem. 2012. Biodiversity loss and its impact on humanity. Nature 486(7401):59–67.

Conrady, M., C. Lampei, O. Bossdorf, W. Durka, and A. Bucharova. 2022. Evolution during seed production for ecological restoration? A molecular analysis of 19 species finds only minor genomic changes. Journal of Applied Ecology 59(5):1383–1393.

Dickens, S.J.M., and K.N. Sudling. 2013. Spanning the science-practice divide: Why restoration scientists need to be more involved with practice. Ecological Restoration 31(2):134–140.

Dockry, M.J., S.J. Hoagland, A.D. Leighton, J.R. Durglo, and A. Pradhananga. 2022. An assessment of American Indian forestry research, information needs, and priorities. Journal of Forestry 1–15. https://doi.org/10.1093/jofore/fvac030.

Doherty, K.D., B.J. Butterfield, and T.E. Wood. 2017. Matching seed to site by climate similarity: Techniques to prioritize plant materials development and use in restoration. Ecological Applications 27(3):1010–1023.

Dumroese, R.K., T. Luna, and T.D. Landis, Eds. 2009. Nursery Manual for Native Plants: A Guide for Tribal Nurseries - Volume 1: Nursery Management. Agriculture Handbook 730. Washington, DC: US Department of Agriculture, US Forest Service.

Eisenberg, C., C. Anderson, A. Collingwood, R. Sissons, C. Dunn, G. Meigs, D. Hibbs, S. Murphy, S. Kuiper, J. SpearChief-Morris, L. Bear, B. Johnston, and C. Edson. 2019. Out of the ashes: Ecological resilience to extreme wildfire, prescribed burns, and indigenous burning in ecosystems. Frontiers in Ecology and Evolution 7:436 doi: 10.3389/fevo.2019.00436.

Ensslin, A., O. Tschöpe, M. Burkart, and J. Joshi. 2015. Fitness decline and adaptation to novel environments in ex situ plant collections: Current knowledge and future perspectives. Biological Conservation 192:394–401.

Espeland, E.K., N.C. Emery, K.L. Mercer, S.A. Woolbright, K.M. Kettenring, P. Gepts, and J.R. Etterson. 2017. Evolution of plant materials for ecological restoration: Insights from the applied and basic literature. Journal of Applied Ecology 54(1):102–115.

Farley, C., T. Ellersick, and C. Jasper. 2015. Forest Service Research and Development Tribal Engagement Roadmap. FS-1043. https://www.fs.usda.gov/research/docs/tribal-engagement/consultation/roadmap.pdf (accessed December 13, 2022).

Gamfeldt, L., T. Snäll, R. Bagchi, M. Jonsson, L. Gustafsson, P. Kjellander, M.C. Ruiz-Jaen, M. Fröberg, J. Stendahl, C.D. Philipson, G. Mikusiński, E. Andersson, B. Westerlund, H. Andrén, F. Moberg, J. Moen, and J. Bengtsson. 2013. Higher levels of multiple ecosystem services are found in forests with more tree species. Nature Communications 4:1340. doi:10.1038/ncomms2328.

Gross, K., B.J. Cardinale, J.W. Fox, A. Gonzalez, M. Loreau, H.W. Polley, P.B. Reich, and J. van Ruijven. 2014. Species richness and temporal stability of biomass production: A new analysis of recent biodiversity experiments. American Naturalist 183:1–12. doi:10.1086/673915.

Havens, K., E.O. Guerrant, M., Maunder, and P. Vitt. 2004. Guidelines for ex situ conservation collection management: Minimizing risks. In E.O. Guerrant, K. Havens, and M. Maunder (Eds.), Ex Situ Plant Conservation: Supporting Species Survival in the Wild (pp. 454–473). Essay, Island Press.

Havens, K., P. Vitt, S. Still, A.T. Kramer, J.B. Fant, and K. Schatz. 2015. Seed sourcing for restoration in an era of climate change. Natural Areas Journal 35(1):122–133.

Hector, A., and R. Bagchi. 2007. Biodiversity and ecosystem multifunctionality. Nature 448:188–190. https://doi.org/10.1038/nature05947.

Hereford, J. 2009. A quantitative survey of local adaptation and fitness trade-offs. American Naturalist 173(5):579–588.

Hooper, D.U., E.C. Adair, B.J. Cardinale, J.E. Byrnes, B.A. Hungate, K.L. Matulich, A. Gonzalez, J.E. Duffy, L. Gamfeldt, and M.I. O'Connor. 2012. A global synthesis reveals biodiversity loss as a major driver of ecosystem change. Nature 486(7401):105–108.

Hornborg, A. 2006. Animism, fetishism, and objectivism as strategies for knowing (or not knowing) the world. Ethnos 71(1):21–32.

Hufford, K.M., and S.J. Mazer. 2003. Plant ecotypes: Genetic differentiation in the age of ecological restoration. Trends in Ecology & Evolution 18(3):147–155.

Husband, B.C., and L.G. Campbell. 2004. Population responses to novel environments: Implications for ex situ plant conservation. In E.O. Guerrant, K. Havens, and M. Maunder (Eds.), Ex Situ Plant Conservation: Supporting Species Survival in the Wild (pp. 231–266). Essay, Island Press.

Johnson, R., L. Stritch, P. Olwell, S. Lambert, M. Horning, and R. Cronn. 2010. What are the best seed sources for ecosystem restoration on BLM and USFS lands? Native Plants Journal 11:117–131.

Johnson, R.C., M.E. Horning, E.K. Espeland, and K. Vance-Borland. 2015. Relating adaptive genetic traits to climate for Sandberg bluegrass from the intermountain western United States. Evolutionary Applications 8(2):172–184.

Kimmerer, R.W. 2013. Braiding Sweetgrass: Indigenous Wisdom, Scientific Knowledge, and The Teachings of Plants. Milkweed Editions: Minneapolis.

Kinzig, A.P., S.W. Pacala, and D. Tilman, Eds. 2002. The Functional Consequences of Biodiversity: Empirical Progress and Theoretical Extensions. Princeton, NJ: Princeton University Press.

Kramer, A.T., and K. Havens. 2009. Plant conservation genetics in a changing world. Trends in Plant Science 14(11):599–607.

Kramer, A.T., S. Wood, S. Frischie, and K. Havens. 2018. Considering ploidy when producing and using mixed-source native plant materials for restoration. Restoration Ecology 26.

Leger, E.A., S. Barga, A.C. Agneray, O. Baughman, R. Burton, and M. Williams. 2021. Selecting native plants for restoration using rapid screening for adaptive traits: Methods and outcomes in a Great Basin case study. Restoration Ecology 29(4):e13260.

Leimu, R., and M. Fischer. 2008. A meta-analysis of local adaptation in plants. PLoS One 3(12):e4010.

Linhart, Y.B., and M.C. Grant. 1996. Evolutionary significance of local genetic differentiation in plants. Annual Review of Ecology and Systematics 27(1):237–277.

Loreau, M., and A. Hector. 2001. Partitioning selection and complementarity in biodiversity experiments. Nature 412:72–76. https://doi.org/10.1038/35083573.

Luna, T., D. Thomas, T.D. Landis, and J. Pinto. 2003. Tribal Nursery Needs Assessment. USDA Forest Service, State and Private Forestry: Reforestation, Nurseries, and Genetic Resources (RNGR), Southern Research Station. https://rngr.net/inc/Tribal%20Nursery%20Needs%20Assessment.pdf (accessed February 10, 2023).

Massatti, R., R.K. Shriver, D.E. Winkler, B.A. Richardson, and J.B. Bradford. 2020. Assessment of population genetics and climatic variability can refine climate-informed seed transfer guidelines. Restoration Ecology 28(3):485–493.

Omernik, J.M. 1987. Ecoregions of the conterminous United States. Annals of the Association of American Geographers 77(1):118–125. doi:10.1111/j.1467-8306.1987.tb00149.x.

Pierotti, R., and D. Wildcat. 2000. Traditional ecological knowledge: The third alternative (Commentary). Ecological Applications 10(5):1333–1340.

Pizza, R., E. Espeland, and J. Etterson. 2021. Eight generations of native seed cultivation reduces plant fitness relative to the wild progenitor population. Evolutionary Applications 14(7):1816–1829.

Prober, S., M. Byrne, E. McLean, D. Steane, B. Potts, R. Vaillancourt, and W. Stock. 2015. Climate-adjusted provenancing: a strategy for climate-resilient ecological restoration. Frontiers in Ecology and Evolution (3). doi:10.3389/fevo.2015.00065.

Saari, C., and W. Glisson. 2012. Survey of Chicago Region Restoration Seed Source Policies. Ecological Restoration 30:162–165.

Shaw, N., and S. Jensen. 2014. The challenge of using native plant materials for sagebrush steppe restoration in the Great Basin, USA (Chapter 4.2). In K. Kiehl, A. Kirmer, N. Shaw, and S. Tischew, eds. Guidelines for Native Seed Production and Grassland Restoration (pp. 141–159). Newcastle upon Tyne, UK: Cambridge Scholars.

Vitt, P., J. Finch, R.S. Barak, A. Braum, S. Frischie, and I. Redlinski. 2022. Seed sourcing strategies for ecological restoration under climate change: A review of the current literature. Frontiers in Conservation Science 3.

Viveiros de Castro, E. 2004. Exchanging perspectives: The transformation of objects into subjects in Amerindian ontologies. Common Knowledge 10(3):463–484.

9

Summary Conclusions and Recommendations for the Native Seed Supply

DISCUSSION

The seeds of native plant species are unique, and their supply is closely tied to the natural landscape from which they originate. At a time when the demand for native seed in ecological restoration and other applications is increasing sharply, many of those landscapes are subject to human-induced environmental stressors that imperil the immense biodiversity they embody. Therefore, native seeds are needed from natural landscapes to improve the ecosystem functionality of degraded landscapes, and for many other applications that benefit from their use.

This is also a time of major transition in the types of demand for native seed, and the needs are evolving. Over the last several decades, there has been a shift from the use of non-native seed species to an increased use of native species in restoration and rehabilitation. A greater diversity of native species of some types, such as grasses, is now available, but far less so for others, such as forbs. For some applications, such as on eroded slopes, and after fires, there is a lack of native seeds that meet short-term performance needs such as rapid soil coverage, so non-native seeds are used.

Public-sector seed buyers at the federal and state level are increasingly requesting native seed sourced from specific geographic locations or that is otherwise considered locally adapted. This demand trend is also happening in restoration projects on private land. The committee learned from seed buyers that the availability of native seed is insufficient, both in terms of its availability when needed, and from the standpoint of the desired diversity of species, ecotypes, geographic origin, and other characteristics. Substitutions for desired species or seeds from desired seed zones are not rare even for routine uses (e.g., maintenance, conservation program uses, habitat creation) and a common occurrence with respect to emergency needs.

On the other hand, the committee learned from suppliers of native seed that uncertain demand is a major barrier to their ability to provide what buyers need. They indicate that they are poised to produce additional seed of much needed diverse species and ecotypes if they have access to the necessary source material to begin production and a clearer and more consistent signal for demand.

As the committee pondered the juxtaposition of how this increasing demand for native seed apparently is not enough to stimulate a sufficient supply response to fill the need, several puzzle pieces began to connect. One large piece of the puzzle is that a large diversity of existing native plant populations is on public lands, and in the West, managed by several federal agencies, with their properties intertwined across almost half of the land area, adjacent to state, tribal land, and private lands. The federal government is far less present in the eastern part of the United States, but many natural areas are under management of state and local governments and by land trusts on

111

private reserves. In some cases, as the committee learned in the case of Iowa, virtually the only remaining natural prairies in the state exist along the sides of public roads. Thus, the source of much of the stock seed that suppliers need to produce primarily exists on publicly managed lands.

The solution to the native seed supply shortage is not to rely solely on the direct use of wild collections for all public projects, which would dangerously decrease the resource. The native plant communities on public land would be better used as the building blocks for the comprehensive assembly of a sustainable, public-private native seed industry. Congress charged the federal land-management agencies to carry out a plan for a native seed supply. The collection of seed from public lands needs to be managed as a critical renewable resource with the goal of building that supply—one that is scoped to be representative of the genetic diversity across the landscape and scaled to meet the nation's growing needs.

The implications to addressing this need include employing sufficient human resources and expertise dedicated to carrying out this comprehensive effort. Additional personnel are needed on multiple fronts, including for the responsibility of monitoring native plant populations across public lands, their overall health, and the genetics they embody, and for oversight of a methodical approach to seed collection based on rigorous protocols. The Seeds of Success program, with more than 27,000 accessions so far, is the example of a coordinated, high-quality collecting and seed banking effort to target wild populations of plant species in geographical locations of interest. To meet the need, that program needs to be conducted under the framework of a larger vision, at a bigger scale and in a wider breadth of geographies nationwide.

The next fundamental step to building the seed supply involves bringing banked seed into field cultivation through the complex transition from conditions in the wild to that of managed agriculture, and providing oversight of the seeds' genetics during increase and multiplication—essentially the process of native seed development—that would result in more than just a single increase for one or two projects, but also in stock seed material for distribution to suppliers to produce seed with transparent origins, and clear identities and attributes.

Who can do this? A precedent for a process like this is currently led by the US Department of Agriculture's Natural Resources Conservation Service (USDA NRCS) Plant Materials Centers (PMCs), which have for many years conducted plant development to feed the seed industry's pipeline for soil conservation programs on private lands. Many public land projects have used seed from the PMCs and continue to do so today. The PMCs maintain lines of stock seed for the private-sector seed industry, and help producers learn how to best grow the seeds economically. That work is complemented by USDA NRCS's technical staff who provide advice to private landowners enrolled in conservation programs on the appropriate seeds to use for different conservation practices. Because the PMCs already supplied the industry with stock seed, these advisers support the demand for conservation seed purchases and direct landowner-buyers to seed types they know are available in the market.

This business model is an important example for consideration, although for several reasons, it does not solve all the problems of the native seed supply. The PMCs have become oriented to meeting primarily private-sector soil conservation needs. They also maintain some seed lines developed using a degree of selection or genetic manipulation for agronomic traits, such as yield, that many would consider inappropriate for restoration, but they also develop natural track plant materials. For example, the PMCs do collaborate with efforts like the not-for-profit Texas Natives Seeds to develop supplies of native seeds (certified, natural track germplasm) that serve both restoration and revegetation goals.

There also does not yet exist a widespread advisory function for buyers of seed for restoration needs that parallels the one that the NRCS provides to landowners on soil conservation programs. Finally, at present, the PMCs are very leanly staffed, so an expansion of both the PMC mission and its capacity would be needed to meet the diverse needs of the federal land-management agencies, states, and private users of the nation.

To make the seed supply more robust, the many information gaps that affect the ability of the native seed supply to function efficiently and effectively need to be filled. Some of the needs are described in Chapter 8, and addressing them would inform buyers' decision making, reduce uncertainty for suppliers, and improve restoration outcomes. Native seed suppliers also need production knowledge about difficult-to-grow species, and suppliers and users would benefit from more consistent seed testing methodologies. Progress is being made, but too slowly, and the resources needed to fill these gaps are insufficient, which holds the seed supply back. In addition to research needs, information about native seed use, prices paid for seed, seed contracting arrangements that share risk between buyers and suppliers, acres restored, outcomes of seeding projects, and species used is not routinely documented and shared by land-management or statistical agencies. Better documentation would facilitate better planning.

The common needs of land-management agencies are a central piece of the puzzle. In addition to oversight for the public lands on which many native plant populations grow, the federal, state, and tribal land-management agencies have similar uses and needs for native seeds, and in many regions, the lands under their control are physically contiguous, so they have common plant biodiversity needs. Generally, however, each agency addresses its seed needs independently and all find it difficult to obtain the supply of seeds needed to fully address their continual needs. Some, like the US Forest Service (USFS), have nurseries and seed-cleaning capacity to address a portion of its needs. The Bureau of Land Management (BLM) has warehouses that provide short-term storage for purchased seed. Other agencies and groups, such as the National Park Service, and most tribal nations, do not have internal units dedicated to planning or tracking native seed needs or purchases, and simply work on a project-by-project basis when funds are available. The committee found that agencies with multiple-year planning horizons, with priorities updated on a yearly basis, such as the Integrated Natural Resource Management Plans used by Department of Defense land managers, appeared to be more successful in obtaining their desired seeds. Why not bring the considerable expertise, infrastructure, best practices, and assets that exist across the federal agencies together to address their common goals?

The committee found that the agencies do engage in collaborations on an ad hoc basis. In response to the 2001 congressional mandate to develop a native seed supply, BLM and the USFS began to establish regional programs for native plant restoration and native plant materials development but have not been able to expand those programs to the geographic scale required to meet the seed needs of the federal agencies, much less the nation. The projects led by these agencies, such as those in the Pacific Northwest, Great Basin, Colorado Plateau, and Mohave Desert, are examples of the productive direction that cooperative action can take, but those are pilot-level efforts. These programs are limited relative to the need, and most regions across the nation lack any similarly focused programs. The National Seed Strategy developed in 2015 by BLM and other federal agencies and non-federal partners conceptually recognized the need for extensive interagency and regional cooperation to meet its stated goals. For now, however, it is largely an aspirational blueprint pointing the way forward, and independent activities on elements of the blueprint comprise the extent of collective action.

The committee concluded that there is much to gain by taking interagency coordination a step further to build a national native seed supply for public lands, and beyond. Such a step necessitates a greater and more sustained level of engagement and funding. The extensive coordination required to address native seed needs is comparable to that employed for wildland fire response, and carried out through the National Interagency Fire Center, a home for the fire management programs of the federal land-management agencies and other federal units that deal with fire. It coordinates wildland firefighting across the United States using a regional approach to engage all levels of federal and state governments, the tribes, and the private sector in planning fire responses, mobilizing emergency resources, conducting research, carrying out data collection and analysis, and communicating policy and best practices.

The same kind of focus and cooperative structure for addressing native seed needs could unify the agencies' independent efforts to meet seed needs for restoration and rehabilitation, if consistent long-term funding was provided to carry out those activities on a large scale. Rather than compete with existing activities within agencies, it could reduce duplication, facilitate sharing of resources, and generate momentum to build the native seed supply that Congress requested more than 20 years ago.

The committee recognizes that bringing separate agencies together in this vision is not a simple task. However, the native seed needs and ecological restoration needs of the country are significant and the current pace at which they are being addressed will be overwhelmed in the coming years as climate change, habitat destruction, and species extinction intensifies. It is an understatement to say that time is of the essence to bank the seeds and the genetic diversity our lands hold for future extensive planting projects that will require large-scale seed production.

Recent legislation to provide funding for federal and state-level restoration efforts signifies that the congressional interest in restoring the nation's natural heritage is a priority. The needs of millions of acres of ecologically impaired landscapes across the nation have largely been unaddressed for lack of funds to conduct proactive restoration activities. At this moment in time there are both the opportunity and the financial resources to act, so it is urgent that a supply of native seeds appropriate to these landscapes be developed. The funds associated with this legislation represent a tremendous opportunity for the native seed supply and for the conservation and restoration of native plant communities. Addressing them proactively is a unique opportunity to build and stabilize the native seed supply chain.

The committee provides the following conclusions that represent a summary of those in the preceding chapters, as noted in the chapter numbers and conclusion numbers at the end of each conclusion. The summary conclusions provide an overarching rationale for the recommendations that follow them. The recommendations are for action-able steps that can be taken to move the nation closer to the important goal of achieving a robust native seed supply.

Conclusion 1.0: *Native plant communities are the source of seed for a diversity of seeding needs. There is urgency to the conservation and restoration of native plant communities and to building a native seed supply, as climate change and biodiversity loss have put many natural landscapes at increasing risk. Seed shortages continue to be a major barrier to restoration. The implementation of federal plans to build a native seed supply for public lands needs to be accelerated. Developing reliable seed supplies for ecological restoration is an achievable goal, but one that demands substantial inter-institutional commitment to work together on a comprehensive vision to support the development of a more robust seed supply industry, and at a much more intensive and expansive level than is currently under way.* (Conclusions 1-2, 1-3, 1-4, 1-5, 3-2, 3-4, and 4-2)

Recommendation 1.0: The leadership of the Departments of the Interior (DOI), Agriculture (USDA), and Defense (DOD) should move quickly to establish an operational structure that facilitates sustained inter-agency coordination of a comprehensive approach to native plant materials development and native plant restoration. An interagency approach focused on fulfilling the long-standing congressional mandate to develop a native seed supply for public lands could unify the agencies' independent efforts to meet seed needs for restoration and rehabilitation. This focused effort would maximize returns on investment in native plant materials development and restoration and would augment, not replace or compete with, existing activities within agencies. One possible model for agency coordination on a national basis, organized regionally, is the National Interagency Fire Center.

Among the types of activities that could be the focus of concerted efforts include the following:

- Serve as the central coordinating platform for implementing the development of a national native seed supply for rehabilitation and restoration.
- Assist the launch, support, and oversight of regional native seed supply development activities.
- Coordinate the prioritization of species and ecotypes to meet seed needs for different regions.
- Co-develop national policy for native seed collection, seed sharing, and seed use.
- Co-develop and share best management practices for seed choice in restoration.
- Coordinate, prioritize, and support basic and applied research such as described in Chapter 8 and other region-specific needs identified in the National Seed Strategy.
- Review and strengthen policy guidance for the use of native seeds on public lands.
- Co-develop adaptive management approaches that use experimentation during restoration, gather data on outcomes, and use these data to guide future restoration actions.
- Provide a national, central data-collection platform and analytical capability.
- Serve as a focal point for training on seed collection protocols, storage practices, seed cleaning and testing, and other technologies.
- Produce informational resources for stakeholders in native seed supplies and native plant restoration.

Conclusion 2.0: *In some regions (at, above, or below the state level), native seed needs are being addressed by networks or partnerships that include federal agencies, states, tribes, local governments, and the private and nonprofit sectors. Further development of these regional networks, plus greater coordination among them, is a promising way to stabilize demand, expand supply, and increase the sharing of information and technology that is critical to meeting native seed needs.* (Conclusions 4-1, 5-1, 5-2, 6-1, 6-2, 6-3)

Recommendation 2.0 Federal land-management agencies should participate in building regional programs and partnerships to promote native plant materials development and native plant restoration, helping to establish such regional programs in areas where they do not yet exist. Ideally, the existing regional programs and partnerships would grow into a complete nationwide network, assisted by the federal interagency coordinating structure envisioned in Recommendation 1.0. The size, geographic coverage, and membership of the regional programs would vary based on regional needs and would include many existing entities such as the USDA Plant Materials Centers and Agricultural Research Service seed banks, other seed banks, botanic gardens, public nurseries, universities, and other organizations. The regional partnerships could take on the following responsibilities:

2.1 Develop seed priorities. Each region has its workhorse native plant species, ones that are easily grown and highly abundant in natural communities, as well as its other priority species such as those that are important for pollinators and wildlife. Developing region-specific lists of priority species will enable suppliers to focus on developing stocks of the species, and ecotypes of these species, that are likeliest to be in high demand.

2.2 Conduct scenario planning and monitor climate change and habitat loss effects. To enable suppliers to anticipate and meet future needs, scenario planning at the regional level can be used to estimate the kinds and quantities of seeds likely to be in demand over multiyear horizons. Such planning scenarios would be based on estimates of current degraded land and future events likely to cause further degradation. These would be periodically updated through ongoing monitoring.

2.3 Collect wildland seed and curate stock seed. Regional programs should conduct or oversee the collection, increase, and curation of wildland-sourced seed to make it available for future production. Seed must be collected using protocols to record source locations, maximize genetic diversity, and protect wild populations. As banked stocked seed accessions deplete, additional wild collections will be necessary. As source populations are lost, or better ones located, ongoing monitoring and reevaluation will be needed.

2.4 Share information. Regional programs should develop informational tools derived from monitoring regional seed use practices and the success or failure of outcomes. This will inform smarter, more predictable selections of species for procurement in the regional marketplace and more effective restoration strategies.

Conclusion 3.0: *Tribal uses for native plants parallel those of federal and state agencies and their needs include seed collection, increase, seed testing, implementation of seed zones, storage, and contracting for ecological restoration. The historic complexity of land management issues and interactions with federal and state governments, former Bureau of Indian Affairs (BIA) policies associated with cultural assimilation, and a lack of resources have constrained all aspects of tribal land management including activities to build capacity to meet native plant needs.* (Conclusions 5-1, 5-2, 5-3)

Recommendation 3.0: The Bureau of Indian Affairs should work with the Inter-Tribal Nursery Council to promote and expand tribal nurseries. The Bureau of Indian Affairs should prioritize and support tribal native plant uses and capacities, consistent with the Bureau's legal and fiduciary obligation to protect tribal treaty rights, lands, assets, and resources held under the Federal-Tribal Trust. Recognizing tribal sovereignty and self-determination, expanding the cultural, economic, and restoration uses of native plants by tribes will require the promotion and expansion of tribal nurseries and greater support for the Inter-Tribal Nursery Council. Additionally, tribal leaders and land managers should be fully engaged in planning, conducting, and applying results from scientific projects related to seed production and conservation, native plant restoration, and ecosystem management on tribal land.

Conclusion 4.0: *Suppliers view unpredictable demand as their leading challenge. Suppliers indicate that seed contracts with clear delivery timelines, price guarantees, and a guarantee to purchase a predetermined quantity of seed of the specified ecotypes are most important to them.* (Conclusions 3-1, 4-3, 6-6, 7-1, 7-2, 7-3, 7-4, 7-5, 7-6, 7-8, and 7-10)

Recommendation 4.0: The public agencies that purchase native seed should assist suppliers by taking steps to reduce uncertainty, share risk, increase the predictability of purchases, and help suppliers obtain stock material.

To stabilize the native seed supply, seed buyers could take the following actions:

4.1 Conduct proactive restoration on a large scale. Millions of acres of US public land are ecologically impaired. With new federal resources for restoration, federal and state agencies should plan restoration projects on a 5-year basis, ensure that stock seed has been made available to suppliers, and set annual purchase targets for the collection and acquisition of needed ecotypes of native plant species. These actions will result in considerable expansion and stabilization of the market for native seeds, benefiting suppliers and users alike.

4.2 Establish clear agency policies on native seed uses. Land-management agencies should establish clear policies on seed use on lands under their stewardship that support the use of locally adapted native plant materials in management activities, along with clearly delimiting the circumstances for allowing exceptions. This will send a strong signal of species and provenance needs to suppliers.

4.3 Support responsible seed collection and long-term seed banking. Intensive and carefully managed seed collection is needed to supply the native plant material enterprise and conserve native plant diversity for the long term.

In cases where native seed is collected from public lands by private suppliers for direct sale and use in restoration, land-management agencies should employ adequate personnel to issue permits and ensure responsible collection.

In other cases, where seed is collected for increase and native plant materials development, the federal agencies should facilitate this activity by extending the Seeds of Success program to include all regions of the United States, and better supporting its activities.

In still other cases, native seed is collected and banked for long-term conservation. To continue building a species-diverse and genetically diverse long-term native seed bank, the federal agencies, led by BLM and the ARS National Plant Germplasm System, should accelerate their collaboration under the Seeds of Success program.

4.4 Contract for seed purchases before production. Public-sector buyers of native seed should amend their policies to enable contracting for purchases before the seed production cycle begins. Forward contracting, in which the buyer agrees to purchase specific seed at a future date, reduces grower risk by ensuring them a market.

4.5 Use marketing contract features that reduce demand uncertainty. Public-sector buyers of native seed should adopt marketing contract features that specify the delivery timeline, guaranteed prices, and guaranteed purchase of predetermined quantities that meet the buyer's specifications.

4.6 Experiment with native seed contract designs. Public-sector buyers of native seed should experiment with cost- and risk-sharing contract designs, such as BLM's Indefinite Delivery Indefinite Quantity (IDIQ) contracts, which are partially supported by a working capital fund.[1] Other approaches also merit investigation, such as Blanket Purchase Agreements. In special circumstances, such as where high-priority species or ecotypes are unavailable, federal agencies should consider issuing contracts to suppliers for research and development.

---

[1] The text of the recommendation was modified after release of the prepublication to clarify that the BLM IDIQ is not intrinsically tied to the working capital fund and can be supported from other funding sources.

4.7 Consider providing premiums for local ecotype use in USDA conservation programs. USDA's Farm Service Agency and state departments of agriculture should explore whether higher co-payments for landowners' use of local ecotypes of native species in conservation program plantings would enhance environmental benefits while supporting a regional native seed industry. A requirement for the use of certified seed would assure growers that their efforts and expenses are not being undercut by seeds of unverified origins.

Conclusion 5.0: *Suppliers indicate that difficult-to-grow species and lack of stock seed from appropriate seed zones or locations are their top two technical challenges. They identified communicating demand (issues related to planning, communication, funding, and the economics of the seed markets) as key to helping them achieve success in providing seeds to buyers.* (Conclusions 1-1, 1-4, 7-2, 7-3, 7-4, 7-8, and 8-3)

Recommendation 5.0: Federal land-management agencies should work with their regional partners to launch an outreach program to provide seed suppliers with critical tools and information. This outreach task should include the following elements:

5.1 Strengthen the role of the National Resources Conservation Services (NRCS) in supporting native seed suppliers. The NRCS Plant Materials Centers (PMCs) can build on their historical role by developing materials and techniques and advising commercial growers on native seed production, emphasizing non-manipulated germplasm and production methods that minimize genetic change.

5.2 Support information sharing on restoration outcomes. Public land-management agencies should share new research and technical knowledge on restoration in publicly available technical progress reports.

5.3 Facilitate communication with growers. To support existing native seed growers and to encourage new ones, federal agencies should provide tools such as an online marketplace, and resources such as workshops and information on propagation, seed cleaning, and other techniques.

Conclusion 6.0: *There are many information gaps that affect the ability of the native seed supply to function efficiently and effectively. Addressing them would inform decision making, reduce uncertainty, and improve restoration outcomes.* (Conclusions 4-4, 5-3, 8-1, 8-2, 8-3, and 8-4)

Recommendation 6.0: The federal government should commit to a rigorous research and development agenda aimed at expanding and improving the use of native seeds in ecological restoration. This includes the following actions:

6.1 Support basic research. Basic research to support restoration in an era of rapid change should be a priority for the National Science Foundation, USDA National Institute of Food and Agriculture, and the US Geological Survey (USGS). Some critical topics include restoration under climate change, species selection to promote ecological function, traditional ecological knowledge, and the economics of the seed market.

6.2 Build technical knowledge. Development and dissemination of new technical knowledge for restoration should be important priorities for the research arms of federal land-management agencies, such as the USDA-ARS, USFS, Department of Defense, and USGS. Priority topics include improving techniques for production, maintenance of genetic integrity and quality, testing, storage, and deployment of native seeds.

6.3 Adaptive management. Public land agencies and their partners should commit to using a rigorous adaptive management approach that documents all features of the restoration plan, uses restoration treatments as experiments to address critical areas of uncertainty, gathers data on outcomes, and uses these data to guide future restoration actions.

6.4 Seed zones. USDA-ARS, USFS, and USGS, in conjunction with regional programs, should seek to develop a more uniform national system of seed zones, which will be an important step in sending clearer signals to both seed suppliers and seed users.

Conclusion 7.0: *An expansion is needed of humidity-controlled seed warehouses across all regions of the United States for temporary storage of seed. To encourage smaller-scale producers to enter the supply chain for native, ecoregional seed of many diverse species, there is a need to expand cooperative seed-cleaning facilities in areas where commercial facilities with specialized equipment capable of handling small lots of native seeds do not exist.* (Conclusions 6-4, 6-5, and 7-7)

Recommendation 7.0: Federal agencies and other public and private partners, including seed suppliers, should collaborate on expanding seed storage and seed-cleaning infrastructure that can be cooperatively cost-shared regionally. Additional storage can improve the availability of seed ready for restoration when urgent but hard-to-predict needs arise. Greater refrigerated and freezer storage availability would protect the viability of purchased seed until its use. Seed-cleaning is also a significant technical challenge for new and small-scale suppliers. Regional native seed-cleaning facilities would encourage more growers to enter the supply chain.

Conclusion 8.0: *BLM is the nation's largest user of native seed for restoration and has the largest capacity for seed storage, but many constraints currently limit its capacity to act as a reliable purchaser of native seeds and thereby to support a more robust native seed industry. The capacity and staffing of its seed warehouses are also inadequate to meet existing and projected needs.* (Conclusion 6-4)

Recommendation 8.0: BLM's Seed Warehouse System needs to be expanded, particularly the capacity for cold storage, and supported by staff with up-to-date knowledge of seed science to manage the seed inventory. The BLM seed warehouses belong in a national program within BLM or the Department of the Interior to ensure they are funded, managed, expanded to meet national needs, and shared among agencies. With more warehouses, seed can be kept near the location where it will be used, reducing transportation costs, time, and seed viability loss.

Conclusion 9.0: *BLM manages extensive areas of natural and seminatural land that constitute an important in-situ repository of seed stock for restoration, but conserving this critical resource is not yet recognized as a land-management objective. The agency's native seed needs depend on the native plant communities that are the source of seeds. There is urgency to the need for conserving the biodiversity that is present in existing native plant communities on BLM land.* (Conclusions 1-3, 1-5, 3-3, and 6-3)

Recommendation 9.0: BLM should identify and conserve locations in which native plant communities provide significant reservoirs of native seeds for restoration. Public land-management agencies should actively recognize and protect the natural plant communities that provide the ultimate sources of native seeds for ecological restoration, using protective designations such as Area of Critical Environmental Concern or Research Natural Areas.

Conclusion 10.0: *Developing reliable seed supplies for ecological restoration is an achievable goal for BLM and other public-sector users of native seed, but one that demands substantial institutional commitment to make restoration a high-priority objective, cultivate and empower botanical and ecological expertise in decision making, and create sustainable funding streams for restoration that enable long-term planning, successful implementation, and learning from experience.* (Conclusions 1-4, 3-1, 3-4, and 8-1)

Recommendation 10.0: The Plant Conservation and Restoration Program (PCRP) should be empowered with additional capacity to plan, oversee restoration, and build stocks of seed. BLM should give greater authority and resources to the PCRP to oversee seed purchasing and warehousing decisions, monitor restoration outcomes, engage in long-term restoration planning, and ensure that staff with plant expertise are available to guide restoration planners and seed purchasers across BLM. This expanded role for PCRP will hasten the pace of change toward the use of natives that is already under way. The PCRP should expand the use of IDIQ or other innovative, risk-sharing contracts to build a diverse supply of native seed in BLM warehouses.

Collectively, these recommendations represent an ambitious agenda for action. The committee believes that agenda is commensurate to the challenges facing our natural landscapes, and to the responsibility of the federal public land-management agencies to take a focused, coordinated leadership role in addressing them. The committee is optimistic that the many public and private parties involved in land stewardship on behalf of tribal nations, states, and other landowners will be willing partners in ensuring their success.

Appendix 1

Committee Biographies

Susan P. Harrison (NAS), *Chair*, is a Distinguished Professor in the Department of Environmental Science and Policy of the University of California, Davis. She was elected to the National Academy of Sciences in 2018 for her excellence in environmental science and ecology. Harrison is a leader in the study of ecological diversity at different spatial and temporal scales, and of the mechanisms and processes that maintain diversity. Her work is of fundamental importance for understanding the impact of global change on ecological communities, and for conservation biology from local to global scales. Dr. Harrison received her B.S. in 1983 in zoology and M.S. in 1986 in ecology from University of California, Davis, and Ph.D. in 1989 in biology from Stanford University.

Delane Atcitty is Executive Director of the Indian Nations Conservation Alliance, an organization that connects underserved native ranchers and farmers to federal agencies and Tribal Leadership that share an interest in conservation and natural resource management. Mr. Atcitty also is the principal of Arrowhead Resource Management, LLC, which provides ranch management and agri-business consulting services to native communities. Previously, he was a Natural Resource Specialist at the Department of the Interior's Bureau of Indian Affairs, a Rangeland Specialist with the Bureau of Land Management, and before that, a Rangeland Specialist for the Nature Conservancy SVR Ranch, overseeing management of cattle and bison on the ranch. Mr. Atcitty is a Board member of the Navajo Agricultural Products Industry, Holistic Management International Board of Director, National Grazing Lands Coalition Advisory Board Member, and Past Chair of the Native American Rangeland Advisory Committee of the Society for Range Management. Mr. Atcitty received a bachelor of applied science (B.A.Sc.) degree in agribusiness in 2007 from Oklahoma Panhandle State University, and in 2009 a masters in ranch management from Texas A&M University's King Ranch Institute.

Rob Fiegener is an independent consultant who previously served as Director of Plant Materials at the Institute for Applied Ecology. Mr. Fiegener led the Institute's native seed collection, production, and distribution activities, including participation in Seeds of Success, the Willamette Valley Native Plant Partnership, and other regional plant materials efforts. He served as director of the Native Seed Network from 2004 to 2020 and produced the series of National Native Seed Conferences from 2010 to 2017. Previously he worked for Oregon State University's Institute for Natural Resources, the US Forest Service, and the National Park Service. He currently serves as chair of the Society for Ecological Restoration's International Network for Seed-based Restoration and is a member of the IUCN Species Survival Commission Seed Conservation Specialist Group. He holds a B.S. in natural resources management from Humboldt State University and a M.S. in ecology from University of California, Davis.

Rachael E. Goodhue is Professor and Department Chair in Agricultural and Resource Economics at University of California, Davis. She earned her Ph.D. and M.S. from University of California, Berkeley, and her B.A. from Swarthmore College. Her research interests include agricultural marketing and organization, agri-environmental policy, pest management, regulation, and contract design. She is on the editorial boards of Agricultural and Food Economics, California Agriculture, and Review of Industrial Organization. Currently she teaches intermediate microeconomic theory for undergraduates in the managerial economics major, and courses on agricultural markets and the economics of California agriculture for graduate students. Goodhue works regularly with the California Department of Food and Agriculture to assess the economic impact of proposed pesticide regulations. Goodhue serves on the California Walnut Board and California Walnut Commission. She is active in university service and served as Chair of the Davis Division of the Academic Senate from 2016 to 2018.

Kayri Havens is Director of Plant Science and Conservation and Senior Scientist at the Chicago Botanic Garden. She is also an adjunct professor of biology at the University of Illinois-Chicago and at Northwestern University. Previously Dr. Havens was a conservation biologist at the Missouri Botanical Garden. Since 1999, she has been a member of the Conservation Committee of the American Public Gardens Association and served as Chair from 2006 to 2008. She is a board member and treasurer of Botanic Gardens Conservation International-US, co-director of the citizen science program Budburst, and a member of the IUCN Seed Conservation Specialist Group. Dr. Havens is also a past president of the Illinois Native Plant Society and founder of the Midwestern Rare Plant Task Force. She has received many awards including the American Horticultural Society Liberty Hyde Bailey Award in 2019 and the Secretary of the Interior's Partners in Conservation Award in 2010. Dr. Havens received her B.A. and M.S. in botany from Southern Illinois University and Ph.D. in biology from Indiana University.

Carol House is an independent consultant who served as a senior program officer for the National Academies of Sciences, Engineering, and Medicine's Committee on National Statistics. Prior to the National Academies, she held several positions at the National Agricultural Statistics Service of the US Department of Agriculture, including deputy administrator for programs and products, associate administrator, director of research and development, and director of survey management. She also served as chair of the Agricultural Statistics Board. She has provided statistical consulting on sample surveys in China, Colombia, the Dominican Republic, and Poland. She is a fellow of the American Statistical Association and an elected member of the International Statistical Institute. Her graduate training was in mathematics at the University of Maryland.

Richard C. Johnson is an adjunct professor at the Regional Plant Introduction Station, Washington State University. He is also a retired Research Agronomist from the US Department of Agriculture Agricultural Research Service where he worked for over 30 years on plant germplasm conservation and utilization. From 2005 to 2016, Johnson led a cooperative program between the National Plant Germplasm System (NPGS) and the BLM Seeds of Success (SOS) program. As a result, more than 10,000 new native plant collections have been acquired for conservation, and thousands of native seed collections have been distributed to private and public entities for research and development. He has also worked extensively with the US Forest Service, BLM, and the Great Basin Native Plant Project for enhancement and utilization of native plant materials. Dr. Johnson has published numerous scientific articles on adaptation of key native species and has developed seed zones to guide the use of germplasm for restoration projects. In 2010 Johnson was a recipient of the US Department of the Interior "Partners in Conservation Award" through Ken Salazar, then Secretary of the Interior. In 2014, he chaired the Interagency Committee to Identify Research Needs and Conduct Research to Provide Genetically Appropriate Seed, and to Improve Technology for Native Seed Production and Ecosystem Restoration. This contributed to the Plant Conservation Alliance's National Seed Strategy for Rehabilitation and Restoration, 2015–2020. Dr. Johnson received B.S. degrees in wildlife biology (1974) and agronomy (1976), and a M.S. in agronomy in 1978, all from Washington State University. In 1981 he received a Ph.D. in agronomy from Kansas State University.

Elizabeth Leger is professor in the Department of Biology at the University of Nevada, Reno (UNR), where she has been faculty since 2006. She received a Ph.D. in plant ecology from the University of California, Davis, and did a post-doc focused on invasive plants at SUNY Stony Brook. Her current research focuses on native plant ecology and restoration in invaded areas of the Great Basin, and she has advised multiple post-doctoral, graduate, and undergraduate students studying the plants of the Great Basin. Dr. Leger served on the UNR faculty senate as the representative for the College of Agriculture, Biotechnology, and Natural Resources for a 3-year term, and served a 3-year term as the associate director for the Ecology, Evolution and Conservation Biology graduate program. She serves the greater scientific community with review and editorial work and has served on multiple grant review panels for federal organizations including the US Department of Agriculture and the National Science Foundation. In addition to her work in plant ecology and restoration, Dr. Leger is the co-creator and director of the University of Nevada, Reno Museum of Natural History, which is a major research, teaching, and outreach institution at UNR.

Virginia Lesser is professor of statistics and Director of the Survey Research Center at Oregon State University. She conducts research on survey methodology, applied statistics, environmental statistics, and ecological monitoring. Dr. Lesser currently works on survey research examining methods to improve response rates through using multiple contact modes, such as Web and mail, and other methods incorporated during administration of the surveys. She has designed and administered over 200 surveys conducted by the Survey Research Center at Oregon State University. She is a fellow of the American Statistical Association and an elected member of the International Statistical Institute. She has served on several National Academies committees, including the Committee on Capitalizing on Science, Technology, and Innovation: An Assessment of the Small Business Innovation Research Program-Phase II; Panel on the Review of the Study Design of the National Children's Study Main Study; Panel on Survey Options for Estimating the Illegal Alien Flow at the Southwest Border; Panel to Review the Occupational Information Network; Committee to Assess the Minerva Research Initiative and the Contribution of Social Science to Addressing Security Concerns; and Committee on the Review of the National Institute of Safety and Health/Bureau of Labor Statistics Respirator Use Survey Program. She has a doctorate in public health in biostatistics from the University of North Carolina, Chapel Hill.

Jean Opsomer is a vice president at Westat in Rockville, Maryland. He was formerly a professor and department chair in the Department of Statistics at Colorado State University, as well as a faculty member at Iowa State University. His research focuses on shape-constrained and nonparametric methods in survey estimation and on several interdisciplinary projects with survey components on a range of topics. He is a member of Statistics Canada's Advisory Committee on Statistical Methods. He previously served on the Bureau of Labor Statistics' Technical Advisory Committee and the USDA's Advisory Committee on Agricultural Statistics. He has served on two National Academies' study committees, including the Panel to Review USDA's Agricultural Resource Management Survey, and the Panel on Improving Data Collection and Reporting about Agriculture with Increasingly Complex Farm Business Structures. Opsomer is an elected fellow of the Institute of Mathematical Statistics and the American Statistical Association, as well as an elected member of the International Statistical Institute. He holds a Ph.D. in statistics from Cornell University.

Nancy Shaw is Emeritus Scientist with the US Forest Service's Rocky Mountain Research Station in Boise, Idaho. Her research over the last 35 years has focused on native plant materials development and restoration of riparian and terrestrial systems in the Intermountain West. From 2003 to 2013, she was Team Leader for the Great Basin Native Plant Project, an interdisciplinary program to develop seed transfer guidelines, seed technology, seed production protocols, and improved methodology for reestablishing native plant communities. The project involved collaboration with 20 federal, state, and private cooperators, including researchers, academics, the native seed industry, practitioners, and students. She is currently a Board member of the Society for Ecological Restoration and Chair-elect for the International Network for Seed-based Restoration, and she served as a member of the Steering Committee for the National Seed Strategy (2014–2015). Awards include the National Grasslands Research Award (USDA FS and NRCS, 2007), National Service First Award (USDA FS, co-awarded, 2007), National Plant Materials Development Award (USDA 4651 FS, 2013), and the Asa Gray Career Achievement Award (USDA FS, 2013). Dr. Shaw holds a Ph.D. in crop science with a seed science emphasis from Oregon State University.

Douglas E. Soltis (NAS) is Distinguished Professor in the Florida Museum and the Department of Biology, University of Florida. Prior to moving to Florida in 2000, he was professor of botany at Washington State University. Research interests include building the tree of all life, flowering plant evolution, and genome doubling (polyploidy). He was born in Sewickley, Pennsylvania. He graduated from the College of William and Mary with a B.A. in biology in 1975. He received his Ph.D. in 1980 from Indiana University. He was named a Distinguished Professor at the University of Florida in 2008. He was president of the Botanical Society of America (1999–2000). He has received the Centennial Award and the Distinguished Fellow Award from the Botanical Society of America. With Pam Soltis, he received the Dahlgren International Prize in Botany (2002) and the Asa Gray Award in Plant Systematics (2006) and Darwin-Wallace Medal (2016). With coauthors P. Soltis, P. Endress, and M. Chase, he received the Stebbins Medal in 2006 (for Phylogeny and Evolution of Angiosperms). He was elected to the American Academy of Arts and Sciences and to the National Academy of Sciences in 2017.

Scott M. Swinton is University Distinguished Professor and Chairperson of the Department of Agricultural, Food, and Resource Economics at Michigan State University (MSU). His economic research looks at agriculture as a managed ecosystem, focusing on management and policy analysis for enhanced ecosystem services. He concentrates on problems involving crop pest, pollination, and nutrient management; precision agriculture; resource conservation; bioenergy crop production; and management of risks to human health and income. Besides his work on US farming, he has extensive experience with agricultural and natural resource management in Latin America and Africa. He teaches undergraduate managerial economics and graduate research design and writing. He is a Fellow in the Agricultural and Applied Economics Association and served as its President in 2018–2019. MSU granted him its William J. Beal Outstanding Faculty award in 2015. Dr. Swinton served on the NAS-IOM Committee on a Framework for Assessing the Health, Environmental, and Social Effects of the Food System and the NAS Committee on Status of Pollinators: Monitoring and Prevention of their Decline in North America. He holds a B.A. from Swarthmore College, a M.S. from Cornell University, and a Ph.D. from the University of Minnesota.

Edward Toth is Director of the Mid-Atlantic Regional Seed Bank (MARS-B). Previously, he founded and directed New York City Parks' Greenbelt Native Plant Center (GNPC) until his retirement in 2021. The GNPC is one of the oldest and largest municipally owned native plant nurseries in the United States, operated in support of conservation and management of the city's natural resources and green infrastructure. In 2012, he initiated the MARS-B, a partner of the national Seeds of Success program, which promotes the use of ecoregionally based seed collection and banking in support of meeting the region's needs for genetically appropriate wild seed. In 2018 he was awarded the Sloan Public Service Award from the Fund for the City of New York. Upon retirement he has incorporated MARSB as an independent nonprofit and is expanding its work throughout the region.

Stanford A. Young is Professor Emeritus of Plant Science at Utah State University (USU) in Logan, Utah. He led the seed certification and foundation seed program at USU and was instrumental in developing native plant germplasm seed certification requirements and standards (both wildland collected and field produced) for the Association of Official Seed Certifying Agencies (AOSCA). He is an Honorary Member of AOSCA. Receiving B.S. and M.S. degrees from USU, Dr. Young attended Oregon State University in Corvallis, Oregon, and earned a Ph.D. in plant pathology and plant breeding in 1977. He worked as a Biochemical Field Specialist for PPG Industries, Inc. based in Fresno, California, and was part-time faculty member at California State University, Fresno, before accepting the position as Seed Certification Specialist in the Plants, Soils, and Climate Department at USU in 1980. Dr. Young served as Chair of AOSCA National Variety Review Boards for Grass and Alfalfa. He was appointed to the Team for DOI-BLM Core Indicators for Monitoring, Plant Materials, and Data Standards in 2011. He is the author of many scholarly publications, newsletters, and bulletins regarding seed certification and seed quality for agronomic crops and native plants. He presently serves as Treasurer for the Great Basin Chapter of the Society for Ecological Restoration.

Appendix 2A

Semi-Structured Interview Guide for Federal Agencies

For any of these questions, if someone else could answer better, please give us the name and contact information for that person(s).

1. Let's begin by describing your position and responsibilities within _____*insert Agency name*_____. Please provide a brief description of your job and the geographic focus of your work.
   Interviewer's prompts if missing from respondent's description:
   - *Position title*
   - *Location of office (city/state)*
   - *Geographic focus of work/responsibility*
   - *How position relates to the purchase/use of native seeds or plants for land management*

Now we are going to focus on native seed and/or plant material purchased and/or used in projects within your scope of responsibility. By "native" we mean seed or plants indigenous to North America prior to European settlement. Plant material would include such things as plants, seedlings and vegetative material, except seed.

2. Do you use any of the following for any purpose on the projects within your scope of responsibility?
   a. Native seed (Y or N)
   b. Non-native seed (Y or N)
   c. Native plant materials (Y or N)
   d. Non-native plant materials (Y or N)
   [If NO to both Q2a and Q2c (no native seed or native plant materials) skip to Q26]

3. For the projects within your scope of responsibility, how often are purchases made of native seed and plant material? Are purchases made on a routine schedule or an "as needed" basis?
   Interviewer's prompt if needed — e.g. several times a year, once a year, once every few years, rarely, etc.
   Interviewer's prompt: Can you share purchase records (3 years) with us?

125

4. For purchases of native <u>seed</u>, thinking over the 3-year period, 2017–2019 (before the COVID-19 pandemic), for projects within your scope of responsibility, what were the most common species of native seed purchased?
Interviewer's note: Here we are only referring to seed and not other plant materials. If the agency provides purchase records, this information may be included in those records. If the respondent reports mixes of native and non-native seed, try to ascertain the most common species of native seed in those mixes.

5. During this same 3-year period, roughly what were the average yearly expenditures for native <u>seed</u>? Your best estimate is fine.
Interviewer's note: Here we are only referring to seed and not other plant materials.
Interviewer's note: If not asked in Q4, ask whether they can share purchase records (3 years). If the respondent reports mixes of native and non-native seed, try to separate the cost of the native seed. If this is not possible, record the expediture for the mix.

6. Now I am going to list a number of potential uses of native seed or plant materials. For each of the following, please indicate whether you used or did not use native seed or plant materials for each purpose during the 3-year period between 2017–2019.

    a. Pollinator habitat projects
      i. Used
      ii. Not Used
      iii. Unsure
    b. Creation/restoration of wildlife habitat (other than pollinator habitat projects)
      i. Used
      ii. Not Used
      iii. Unsure
    c. Restorative activity on land in a wilderness/natural area
      i. Used
      ii. Not Used
      iii. Unsure
    d. Invasive species suppression
      i. Used
      ii. Not Used
      iii. Unsure
    e. Natural disaster recovery from such events as a hurricane, flood, fire, severe drought, etc.
      i. Used
      ii. Not Used
      iii. Unsure
    f. Roadside seeding (after construction)
      i. Used
      ii. Not Used
      iii. Unsure
    g. Roadside maintenance
      i. Used
      ii. Not Used
      iii. Unsure
    h. Stream erosion mitigation/restoration
      i. Used
      ii. Not Used
      iii. Unsure

    i. Soil protection (conservation plantings)
      i. Used
      ii. Not Used
      iii. Unsure
    j. Landscaping (around structures and in parks)
      i. Used
      ii. Not Used
      iii. Unsure
    k. Green infrastructure (stormwater management, etc.)
      i. Used
      ii. Not Used
      iii. Unsure
    l. Rangeland grazing
      i. Used
      ii. Not Used
      iii. Unsure
    m. Energy development remediation (example coalmine, oil drilling, fracking, etc.)
      i. Used
      ii. Not Used
      iii. Unsure
    n. Green strips or vegetative fuel breaks to mitigate wildfire spread
      i. Used
      ii. Not Used
      iii. Unsure
    o. Another purpose, please specify _____

Interviewer note: If the respondent thinks a particular usage fits in more than one category (for example – h) stream erosion may be part of e) natural disaster recovery – then code both uses.

What would you say was the most common use?

7. During this same 3-year period, did any of the procurements specifically seek native seed appropriate for use in a specific local area, or from a specific source location or seed zone.
    a. Yes
    What is your best estimate of the percent of those seeds that were actually purchased versus substitutions that did not fully meet these locality specifications?
    b. No
    Interviewer's note: Here we are trying to determine whether the intended location of seed usage was part of the procurement process. If so, were they actually able to purchase the seed they sought, or was a substitution made?

8. For purchases of native plant material (other than seed), thinking over the 3-year period, 2017–2019, for projects within your scope of responsibility, what were the most common species of native plant materials purchased?
    Interviewer's note: Here we are only referring to plant materials and not seed.
    Interviewer prompt: Can you share purchase records (3 years) with us?

9. During this same 3-year period, roughly what were the average yearly expenditures for these native plant materials? Your best estimate is fine.
    Interviewer's note: Here we are only referring to plant materials and not seed.
    Interviewer prompt: Can you share purchase records (3 years) with us?

10. I am going to list several ways to communicate to suppliers your <u>future needs</u> for native seed. Please indicate which ones you use.
   a. Requests for proposals
      Use
      Not Use
      Unsure
   b. General publicity around projects
      Use
      Not Use
      Unsure
   c. Public meetings with suppliers
      i. Used
      ii. Not Used
      iii. Unsure
   d. Conferences or other professional meetings
      Use
      Not Use
      Unsure
   e. Informal communication with suppliers
      Use
      Not Use
      Unsure
   f. Other, please specify _____

11. In order to evaluate projects that involve native seed or plant materials within your scope of responsibility, do you check on the survival of the seed or plant materials after planting?
   a. Yes, for most project
   b. Yes, for some projects
   c. No → Skip Q12

12. Generally, when is the last time you typically check on the survival of the seed or plant materials after planting?
   a. About 1 year or less after seeding or planting
   b. 1 to 3 years after seeding or planting
   c. More than 3 years after seeding or planting

It is important for the Committee to better understand the decision-making process when [AGENCY NAME] purchases native seed. For these next questions we request details on a "typical project" in which you have been involved.

13. We would like you to identify a typical project involving <u>native seed</u> use in which you have played a fairly major part in the decision-making about purchasing those seed — what species to buy, when to buy, from whom to buy, etc. What is the project's name and can you provide a brief description?
   Interviewer prompt if needed about general description of project, ask more about:
   • *Purpose of project (and what makes it "typical")*
   • *Approximate amount and type of native seed (species, certification, etc.) Acres covered.*
   • *Geographical area of use*
   • *If the project involves both native seed and plant material, include both.*
   • *Regulatory framework – whether regulations or policy is required or encouraged the use native seed and/or locally sourced native seed on the project.*

14. For this specific "typical" project, how was it determined exactly how much and which type of seed (species, germplasm within species, geographic identity, etc.) and plant materials were needed for this project?

15. From what source(s) did you obtain information about the availability of native seed (and plant materials) needed for this project?
    a. In-house knowledge
       Use
       Not Use
       Unsure
    b. Advertising
       Use
       Not Use
       Unsure
    c. Preapproved vendors
       Use
       Not Use
       Unsure
    d. Request for proposals
       Use
       Not Use
       Unsure
    e. Other, please specify _____

16. Please describe the approval and funding process for this project.
    Interviewer prompt if needed –
    • *Did you determine what was needed and then ask for that funding? Did you get the funding you asked for?*
    • *Was the budget for the project determined in another part of your agency, causing you to modify the project specifications to match the budget?*
    • *Were you given an annual budget and had to work within it for multiple projects?*
    • *Were funds leveraged through special programs or other agencies or sources of funding such as partnerships or collaborations?*
    • *What was the specific source of funds for this project?*
    • *Is this process typical for other projects as well?*

17. To purchase the seed used in this project:
    a. Were your seed requirements included in a consolidated buy with other projects in your agency?
    b. Were seeds for this project purchased separately from those needed for other projects?
    c. Other, specify (Example, seed was obtained from stored seed in agency warehouse.)

18. What type of contracting method was used to purchase the seed? Common methods might include purchase directly from a spot market sale, use of a marketing contract ahead of delivery date, or use of a production contract in which there is a sharing of production costs or risks.
    Interviewer prompt if needed – types of contracting arrangements.
    • *Spot market sale of available seed or plant materials from the supplier's current inventory. For example, a bid on a BLM Consolidated Seed Buy. Price and quantity are unknown in advance of the production process.*
    • *Marketing contract that specifies seed or plant materials type, price, quantity, and delivery date. Approved in advance of the sale, this contract may or may not be signed before production decisions have been made or seed is available.*

- *Production contract that specifies seed or plant materials type, desired quantity and delivery date. Contract shares some production costs and/or production risk (by providing flexibility on the quantity delivered and/or the delivery date), such as IDIQ. It may or may not specify a price.*
- *Buying from agency warehouse*
- *Other contracting arrangement – please describe*

19. What input did you have in deciding which supplier was picked?
    Interviewer prompt if needed?
    - *You made the final decision*
    - *You had input but not the final decision*
    - *Decision was made elsewhere, you had little input*

20. I want to understand the timeline for this project. I am going to list six activities and ask for when they took place. Reporting month and year would be helpful.
    a. When did the planning for the project commence?
    b. When was the decision made on what seed were needed?
    c. When was funding authorization finalized?
    d. When were seeds purchased or collection/production contracts awarded?
    e. When were the seeds delivered?
    f. When were the seeds actually utilized on the land?

21. Did project developers reach out to suppliers during the planning stage for technical information and/or information on availability?

22. Did this project eventually receive the amount and type of native seed (plant material) that was in your original "needs assessment?"
    a. Yes
    b. No. What occurred?
       ◦ *Was less seed purchased than requested? Why?*
       ◦ *Was non-native seed substituted for native seed? If yes, because natives were too expensive, natives not available within project timeline, or just not available in general?*
       ◦ *Was native seed having different characteristics substituted for preferred native seed? If yes, because preferred natives were too expensive, preferred natives not available within project timeline, or just not available in general?*

23. In order to evaluate this project, did you check on the survival of the seed after seeding or planting?
    a. Yes
       i. About 1 year or less after seeding or planting
       ii. 1 to 3 years after seeding or planting
       iii. Greater than 3 years after seeding or planting
    b. No

24. How successful was the project in reaching its goals? Was the evaluation useful in setting parameters for similar future projects?

25. What was your overall satisfaction with the decision-making process and availability of native seed for this project? Why?

26. Do you anticipate that native seed and plant material purchase and use on projects within your scope of responsibility is likely to increase or decrease in the short term? . . .in the longer term? Explain:

END. Thank you for your help!

Appendix 2B

State Government Departments Survey Invitation Letter

The National Academies of | SCIENCES
ENGINEERING
MEDICINE

May 24, 2021

<AGENCY NAME>
<DIR NAME>, <TITLE>

The National Academies of Sciences, Engineering, and Medicine is conducting an important assessment of the challenges faced by state government agencies that use native seed and plant materials. I am writing because it is my understanding that your agency uses native seed or plant materials for ecological restoration or other purposes. The assessment will benefit agencies, like yours, by identifying ways for improving connections between supply and demand for native seed and plant materials.

We would greatly appreciate your asking the person who is <u>most knowledgeable</u> about the use of native seed and plant materials at the [Agency Name] to complete this brief online survey that is being sent to state agencies throughout the United States. The survey will ask about how the [Agency Name] uses native seed and plant materials, and the challenges encountered with obtaining native seed.

The online survey is located here: https://opinion.wsu.edu/StateSeedNeeds
Your survey access code is: **<RespID>**

Responses to this survey are voluntary and your answers will be kept completely confidential by the National Academies and the Washington State University Social and Economic Sciences Research Center that is implementing the survey for us. No individual or agency/department will ever be identified in our results. For this study to be successful it is vitally important that we hear from as many agencies as possible.

If you have any questions about why this survey is being done, please contact Krisztina Marton at the National Academies at KMarton@nas.edu. If you have questions about accessing the survey, please contact Thom Allen at Washington State University, ted@wsu.edu, or call him toll-free at 1-800-833-0867.

Thank you for considering our request and we hope to hear from you soon.

Sincerely,

Robin Schoen

Robin Schoen
Director, Board on Agriculture and Natural Resources
The National Academies of Sciences, Engineering, and Medicine

Appendix 2C

State Government Departments Web Survey Instrument

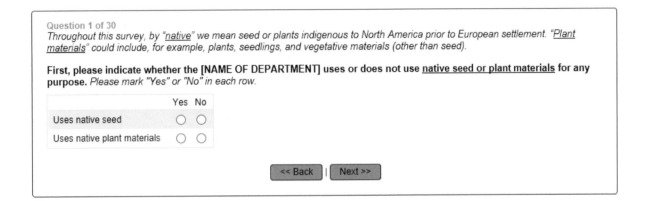

Throughout this survey, by "native" we mean seed or plants indigenous to North America prior to European settlement. "Plant materials" could include, for example, plants, seedlings, and vegetative materials (other than seed).

First, please indicate whether the [NAME OF DEPARTMENT] uses or does not use <u>native seed or plant materials</u> for any purpose. *Please mark "Yes" or "No" in each row.*

	Yes	No
Uses native seed	○	○
Uses native plant materials	○	○

<< Back | Next >>

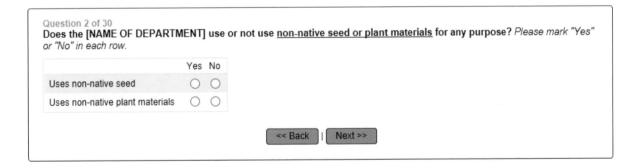

Does the [NAME OF DEPARTMENT] use or not use <u>non-native seed or plant materials</u> for any purpose? *Please mark "Yes" or "No" in each row.*

	Yes	No
Uses non-native seed	○	○
Uses non-native plant materials	○	○

<< Back | Next >>

**[BRANCH: IF NO TO BOTH USES NATIVE SEED AND NATIVE
PLANTS IN Q1 THEN SKIP TO Q27]**

133

Question 3 of 30

For each of the following, please indicate whether the [NAME OF DEPARTMENT] conducts projects that use seed and plant materials at this level? *Please mark "Yes" or "No" in each row.*

	Yes	No
Local or municipal level	○	○
Sub-state regions covering more than one county or municipality	○	○
Statewide	○	○
Other levels, please specify:	○	○

<< Back | Next >>

Question 4 of 30

Please tell us whether you personally have or do not have an active role in the purchase or use of seed or plant materials in the following project areas? *Please mark "Yes" or "No" in each row.*

	Yes, active role	No, no active role
Project planning	○	○
Providing biological expertise	○	○
Purchase execution or contracting	○	○
Project management	○	○
Fieldwork, such as site preparation, seeding, and monitoring	○	○

<< Back | Next >>

Question 5 of 30

During the 3-year period between 2017-2019, which of the following ways were used to obtain native seed or plant materials and which were not used? *Please mark one answer in each row.*

	Used	Not used	Unsure
Purchased from commercial suppliers	○	○	○
Collected or grown by [NAME OF DEPARTMENT]	○	○	○
Obtained from project collaborators	○	○	○
Obtained from any other sources	○	○	○

<< Back | Next >>

[BRANCH: IF <u>NOT USED OR UNSURE</u> TO 5a (COMMERCIAL SUPPLIER) GO TO Q7]

Question 6 of 30

For the native seed or plant materials purchased from <u>commercial suppliers</u> between 2017-2019, which of the following forms of purchase were used and which were not used? *Please mark one answer in each row.*

	Used	Not used	Unsure
Purchased directly on the open market	○	○	○
Purchased via a formal bidding process	○	○	○

<< Back | Next >>

Question 7 of 30

Were any of the native seeds used between 2017-2019 wild collected on state land?

○ Yes

○ No

○ Unsure

<< Back | Next >>

[BRANCH: IF <u>NO OR UNSURE </u>TO Q07 GO TO Q09]

Question 8 of 30

How did the [NAME OF DEPARTMENT] use the native seed that was wild collected on state land between 2017 and 2019? *Please mark one answer in each row.*

	Yes	No	Unsure
The seed was used to plant directly at a project site.	○	○	○
The seed was used to grow plants with the purpose of harvesting seed or plant materials for future use.	○	○	○
The seed was used for another purpose. Please specify:	○	○	○

<< Back | Next >>

Question 9 of 30

In obtaining seed for the [NAME OF DEPARTMENT] projects during 2017-2019, how important or unimportant was it to obtain the following general types of seed? *Please mark one answer in each row.*

	Very important	Somewhat important	Not important	Unsure
Native seed	○	○	○	○
Certified native seed	○	○	○	○
Native seed that was collected from a designated geographic area or specified seed zone	○	○	○	○

<< Back | Next >>

Question 10 of 30

Were there any other important characteristics for the seed that was obtained during the 3-year period between 2017-2019?

○ Yes, please specify: [   ]

○ No

<< Back | Next >>

Question 11 of 30

Were native seed or plant materials used or not used for these purposes during the 3-year period between 2017-2019? *Please mark one answer in each row.*

	Used	Not used	Unsure
Pollinator habitat projects	○	○	○
Creation or restoration of wildlife habitat other than pollinator habitat projects	○	○	○
Restorative activity on land in a wilderness or natural area	○	○	○
Invasive species suppression	○	○	○
Natural disaster recovery from events such as a hurricane, flood, fire or severe drought	○	○	○
Roadside seeding, for example, after construction	○	○	○
Roadside maintenance	○	○	○
Stream erosion mitigation or restoration	○	○	○
Soil protection, such as conservation plantings	○	○	○
Landscaping, for instance such as around structures and in parks	○	○	○
Green infrastructure, such as stormwater management, for example	○	○	○
Another purpose, please specify: [ ]	○	○	○

<< Back | Next >>

Does the [NAME OF DEPARTMENT] have programs to assist private landowners with the use of native seed or plant materials?

○ Yes

○ No

○ Unsure

<< Back | Next >>

Does the [NAME OF DEPARTMENT] contract out seeding or planting activities with native seed or plant materials?

○ Yes

○ No

○ Unsure

<< Back | Next >>

Does the [NAME OF DEPARTMENT] have the following forms of seed storage? *Please mark one answer in each row.*

	Yes	No	Unsure
Ambient storage	○	○	○
Refrigerated storage	○	○	○
Freezer storage	○	○	○

<< Back | Next >>

[BRANCH: IF Q5A (PURCHASED FROM A COMMERCIAL SUPPLIER) IS NO OR UNSURE SKIP TO Q17]

Does the [NAME OF DEPARTMENT] use any of the following contracts? *Please mark one answer in each row.*

	Yes	No	Unsure
Marketing contracts that specifies the type, price, quantity, and delivery date of seed or plant materials? Click here for a definition.	○	○	○
Production contracts that specifies the desired type, quantity and delivery date of native seed and plant materials? It may or may not specify a price. Click here for a definition.	○	○	○

<< Back | Next >>

[BRANCH: IF BOTH Q15a AND Q15b ARE "NO" OR "UNSURE" SKIP TO Q17]

Question 16 of 30
Are the contracts usually established . . .

    ○ before the supplier begins the seed production process?

    ○ after the supplier begins the seed production process?

    ○ unsure of timing of contract

                        [<< Back] | [Next >>]

Question 17 of 30
For each of the following, please indicate whether you use or do not use these sources to obtain information about the availability of native seed and plant materials? *Please mark one answer in each row.*

	Used	Not used	Unsure
In-house knowledge	○	○	○
Advertising	○	○	○
Preapproved vendors	○	○	○
Request for proposals	○	○	○
Other, please specify:	○	○	○

                        [<< Back] | [Next >>]

Question 18 of 30
For each of the following, indicate whether you use or do not use these methods to communicate to suppliers about your anticipated future need for native seed or plant materials? *Please mark one answer in each row.*

	Used	Not used	Unsure
Requests for proposals	○	○	○
General publicity around projects	○	○	○
Conferences or other professional meetings	○	○	○
Informal meetings with growers	○	○	○
Word of mouth	○	○	○
Other, please specify:	○	○	○

                        [<< Back] | [Next >>]

Question 19 of 30
For each of the the following specifications or guidelines, indicate whether they apply or do not apply to the [NAME OF DEPARTMENT] projects that involve native seed or plant materials? *Please mark one answer in each row.*

	Apply	Do not apply	Unsure
Technical specifications	○	○	○
Federal regulations, guidelines, or policies	○	○	○
State regulations, guidelines, or policies	○	○	○
Funding source specifications	○	○	○

<< Back | Next >>

Question 20 of 30
The next few questions ask about any substitution made when the native seed or plant materials sought were unavailable.

How often does the [NAME OF DEPARTMENT] substitute <u>non-native</u> seed or plant materials when native seed or plants are unavailable?

○ Frequently

○ Infrequently

○ Never

<< Back | Next >>

[BRANCH: IF NEVER TO Q20 SKIP TO Q22]

Question 21 of 30
For which of the following reasons were the substitutions with non-native seed or plant materials typically made? *Please mark one answer in each row.*

	Yes	No	Unsure
Because native seed or plant material were available but too expensive.	○	○	○
Because native seed or plant material were available but not within the timeline needed.	○	○	○
Because native seed or plant material were unavailable regardless of price and timing.	○	○	○
Other reasons for the substitutions. Please specify:	○	○	○

<< Back | Next >>

Question 22 of 30

How often does your agency substitute native seed or plant materials <u>having different characteristics</u> when your preferred natives are not available? *Please mark one answer in each row.*

	Frequently	Infrequently	Never
Native seed or plant material of different species	○	○	○
Native seed or plant material from a different region	○	○	○
Other, please specify:	○	○	○

<< Back | Next >>

[BRANCH: IF "NEVER" TO ALL SKIP TO Q24]

Question 23 of 30

For which of the following reasons were the substitutions with <u>native seed or plant materials having different characteristics</u> typically made? *Please mark one answer in each row.*

	Yes	No	Unsure
Because native seed or plant material with the preferred characteristics were available but too expensive.	○	○	○
Because native seed or plant material with the preferred characteristics were available but not within the timeline needed.	○	○	○
Because native seed or plant material with the preferred characteristics were unavailable regardless of price and timing.	○	○	○
Other reasons for the substitutions. Please specify:	○	○	○

<< Back | Next >>

Question 24 of 30

In order to evaluate the [NAME OF DEPARTMENT] projects that involve native seed or plant materials, do you check on the survival of the seed or plant materials after planting?

○ Yes, for most projects

○ Yes, for some projects

○ No

<< Back | Next >>

[BRANCH: IF "NO" SKIP TO Q26]

Question 25 of 30
For a typical project, when is the last time you typically check on the survival of the seed or plant materials after planting?

○ About 1 year or less after seeding or planting

○ 1 to 3 years after seeding or planting

○ More than 3 years after seeding or planting

<< Back | Next >>

Question 26 of 30
Does the [NAME OF DEPARTMENT] have the range of in-house expertise needed for the various aspects of projects that use native seeds or plants, or are there some gaps in the expertise available in-house?

○ We have the expertise we need in-house

○ We have some gaps in expertise. Pease specify:

○ Don't know

<< Back | Next >>

Question 27 of 30
What are the biggest barriers or disincentives to using native seed and plant materials in your work?

<< Back | Next >>

Question 28 of 30
Do you anticipate that the use of native seed or plant materials by the [NAME OF DEPARTMENT] is likely to increase or decrease in the near-term future?

○ Increase

○ Decrease

○ Unsure

Please explain why:

<< Back | Next >>

Question 29 of 30

On the other hand, do you anticipate that the use of native seed or plant materials by the [NAME OF DEPARTMENT] is likely to increase or decrease in the <u>long-term</u> future?

○ Increase

○ Decrease

○ Unsure

Please explain why:

[   ]

<< Back | Next >>

Question 30 of 30

Thinking about the 3-year period between 2017-2019, what was the approximate average annual expenditure on native seed and plant materials combined for the [NAME OF DEPARTMENT]? *Your best estimate is fine.*

$ [   ]   Approximate annual expenditures

<< Back | Next >>

If you have any additional comments, please feel free to type them in the space provided below.

[   ]

Thank you for your help!

<< Back | Next >>

Appendix 2D

State Government Departments Survey Frequency Distributions

Q01A Does [DEPARTMENT] use native seed for any purpose?

		Frequency	Percent	Valid Percent	Cumulative Percent
Valid	Yes	89	91.8	94.7	94.7
	No	5	5.2	5.3	100.0
	Total	94	96.9	100.0	
Missing	No answer	3	3.1		
Total		97	100.0		

Q01B Does [DEPARTMENT] use native plant materials for any purpose?

		Frequency	Percent	Valid Percent	Cumulative Percent
Valid	Yes	85	87.6	94.4	94.4
	No	5	5.2	5.6	100.0
	Total	90	92.8	100.0	
Missing	No answer	7	7.2		
Total		97	100.0		

Q02A Does [DEPARTMENT] use non-native seed for any purpose?

		Frequency	Percent	Valid Percent	Cumulative Percent
Valid	Yes	69	71.1	74.2	74.2
	No	24	24.7	25.8	100.0
	Total	93	95.9	100.0	
Missing	No answer	4	4.1		
Total		97	100.0		

Q02B Does [DEPARTMENT] use non-native plant materials for any purpose?

		Frequency	Percent	Valid Percent	Cumulative Percent
Valid	Yes	60	61.9	65.2	65.2
	No	32	33.0	34.8	100.0
	Total	92	94.8	100.0	
Missing	No answer	5	5.2		
Total		97	100.0		

Q03A Does [DEPARTMENT] conduct projects that use seed and plant materials at this level? Local or municipal level.

		Frequency	Percent	Valid Percent	Cumulative Percent
Valid	Yes	64	66.0	80.0	80.0
	No	16	16.5	20.0	100.0
	Total	80	82.5	100.0	
Missing	Branching skip	3	3.1		
	Partial breakoff	1	1.0		
	No answer	13	13.4		
	Total	17	17.5		
Total		97	100.0		

Q03B Does [DEPARTMENT] conduct projects that use seed and plant materials at this level? Sub-state region covering more than one county or municipality.

		Frequency	Percent	Valid Percent	Cumulative Percent
Valid	Yes	66	68.0	83.5	83.5
	No	13	13.4	16.5	100.0
	Total	79	81.4	100.0	
Missing	Branching skip	3	3.1		
	Partial breakoff	1	1.0		
	No answer	14	14.4		
	Total	18	18.6		
Total		97	100.0		

Q03C Does [DEPARTMENT] conduct projects that use seed and plant materials at this level? Statewide.

		Frequency	Percent	Valid Percent	Cumulative Percent
Valid	Yes	76	78.4	88.4	88.4
	No	10	10.3	11.6	100.0
	Total	86	88.7	100.0	
Missing	Branching skip	3	3.1		
	Partial breakoff	1	1.0		
	No answer	7	7.2		
	Total	11	11.3		
Total		97	100.0		

Q03D Does [DEPARTMENT] conduct projects that use seed and plant materials at this level? Other levels, please specify.

		Frequency	Percent	Valid Percent	Cumulative Percent
Valid	Yes	12	12.4	38.7	38.7
	No	19	19.6	61.3	100.0
	Total	31	32.0	100.0	
Missing	Branching skip	3	3.1		
	Partial breakoff	1	1.0		
	No answer	62	63.9		
	Total	66	68.0		
Total		97	100.0		

Q04A Do you personally have an active role related to the purchase or use of seed or plant materials in this project area? Project planning.

		Frequency	Percent	Valid Percent	Cumulative Percent
Valid	Yes, active role	76	78.4	86.4	86.4
	No, no active role	12	12.4	13.6	100.0
	Total	88	90.7	100.0	
Missing	Branching skip	3	3.1		
	Partial breakoff	2	2.1		
	No answer	4	4.1		
	Total	9	9.3		
Total		97	100.0		

Q04B Do you personally have an active role related to the purchase or use of seed or plant materials in this project area? Providing biological expertise.

		Frequency	Percent	Valid Percent	Cumulative Percent
Valid	Yes, active role	73	75.3	83.0	83.0
	No, no active role	15	15.5	17.0	100.0
	Total	88	90.7	100.0	
Missing	Branching skip	3	3.1		
	Partial breakoff	2	2.1		
	No answer	4	4.1		
	Total	9	9.3		
Total		97	100.0		

Q04C Do you personally have an active role related to the purchase or use of seed or plant materials in this project area? Purchase execution or contracting.

		Frequency	Percent	Valid Percent	Cumulative Percent
Valid	Yes, active role	60	61.9	69.0	69.0
	No, no active role	27	27.8	31.0	100.0
	Total	87	89.7	100.0	
Missing	Branching skip	3	3.1		
	Partial breakoff	2	2.1		
	No answer	5	5.2		
	Total	10	10.3		
Total		97	100.0		

Q04D Do you personally have an active role related to the purchase or use of seed or plant materials in this project area? Project management.

		Frequency	Percent	Valid Percent	Cumulative Percent
Valid	Yes, active role	70	72.2	79.5	79.5
	No, no active role	18	18.6	20.5	100.0
	Total	88	90.7	100.0	
Missing	Branching skip	3	3.1		
	Partial breakoff	2	2.1		
	No answer	4	4.1		
	Total	9	9.3		
Total		97	100.0		

Q04E Do you personally have an active role related to the purchase or use of seed or plant materials in this project area? Fieldwork, such as site preparation, seeding, and monitoring.

		Frequency	Percent	Valid Percent	Cumulative Percent
Valid	Yes, active role	60	61.9	68.2	68.2
	No, no active role	28	28.9	31.8	100.0
	Total	88	90.7	100.0	
Missing	Branching skip	3	3.1		
	Partial breakoff	2	2.1		
	No answer	4	4.1		
	Total	9	9.3		
Total		97	100.0		

Q05A Between 2017-2019 did you obtain native seed or plant materials in this way? Purchased from commercial suppliers.

		Frequency	Percent	Valid Percent	Cumulative Percent
Valid	Used	85	87.6	95.5	95.5
	Not used	2	2.1	2.2	97.8
	Unsure	2	2.1	2.2	100.0
	Total	89	91.8	100.0	
Missing	Branching skip	3	3.1		
	Partial breakoff	3	3.1		
	No answer	2	2.1		
	Total	8	8.2		
Total		97	100.0		

Q05B Between 2017-2019 did you obtain native seed or plant materials in this way? Collected or grown by you.

		Frequency	Percent	Valid Percent	Cumulative Percent
Valid	Used	43	44.3	48.9	48.9
	Not used	39	40.2	44.3	93.2
	Unsure	6	6.2	6.8	100.0
	Total	88	90.7	100.0	
Missing	Branching skip	3	3.1		
	Partial breakoff	3	3.1		
	No answer	3	3.1		
	Total	9	9.3		
Total		97	100.0		

Q05C Between 2017-2019 did you obtain native seed or plant materials in this way? Obtained from project collaborator.

		Frequency	Percent	Valid Percent	Cumulative Percent
Valid	Used	55	56.7	62.5	62.5
	Not used	24	24.7	27.3	89.8
	Unsure	9	9.3	10.2	100.0
	Total	88	90.7	100.0	
Missing	Branching skip	3	3.1		
	Partial breakoff	3	3.1		
	No answer	3	3.1		
	Total	9	9.3		
Total		97	100.0		

Q05D Between 2017-2019 did you obtain native seed or plant materials in this way? Obtained from any other sources.

		Frequency	Percent	Valid Percent	Cumulative Percent
Valid	Used	26	26.8	32.1	32.1
	Not used	27	27.8	33.3	65.4
	Unsure	28	28.9	34.6	100.0
	Total	81	83.5	100.0	
Missing	Branching skip	3	3.1		
	Partial breakoff	3	3.1		
	No answer	10	10.3		
	Total	16	16.5		
Total		97	100.0		

Q06A Between 2017-2019 did you use this way of purchasing native seed or plant materials from commercial suppliers? Purchased directly on the open market.

		Frequency	Percent	Valid Percent	Cumulative Percent
Valid	Used	66	68.0	80.5	80.5
	Not used	8	8.2	9.8	90.2
	Unsure	8	8.2	9.8	100.0
	Total	82	84.5	100.0	
Missing	Branching skip	7	7.2		
	Partial breakoff	4	4.1		
	No answer	4	4.1		
	Total	15	15.5		
Total		97	100.0		

Q06B Between 2017-2019 did you use this way of purchasing native seed or plant materials from commercial suppliers? Purchased via a formal bidding process.

		Frequency	Percent	Valid Percent	Cumulative Percent
Valid	Yes	52	53.6	65.8	65.8
	No	15	15.5	19.0	84.8
	Unsure	12	12.4	15.2	100.0
	Total	79	81.4	100.0	
Missing	Branching skip	7	7.2		
	Partial breakoff	4	4.1		
	No answer	7	7.2		
	Total	18	18.6		
Total		97	100.0		

Q07 Were any of the native seeds used between 2017-2019 wild collected on state land?

		Frequency	Percent	Valid Percent	Cumulative Percent
Valid	Yes	38	39.2	44.2	44.2
	No	23	23.7	26.7	70.9
	Unsure	25	25.8	29.1	100.0
	Total	86	88.7	100.0	
Missing	Branching skip	3	3.1		
	Partial breakoff	5	5.2		
	No answer	3	3.1		
	Total	11	11.3		
Total		97	100.0		

Q08A How did the [Department] use the native seed that was wild collected on state land between 2017 and 2019? The seed was used to plant directly at a project site.

		Frequency	Percent	Valid Percent	Cumulative Percent
Valid	Yes	36	37.1	92.3	92.3
	No	2	2.1	5.1	97.4
	Unsure	1	1.0	2.6	100.0
	Total	39	40.2	100.0	
Missing	Branching skip	51	52.6		
	Partial breakoff	5	5.2		
	No answer	2	2.1		
	Total	58	59.8		
Total		97	100.0		

Q08B How did the [Department] use the native seed that was wild collected on state land between 2017 and 2019? The seed was used to grow plants with the purpose of harvesting seed or plant materials for future use.

		Frequency	Percent	Valid Percent	Cumulative Percent
Valid	Yes	30	30.9	78.9	78.9
	No	7	7.2	18.4	97.4
	Unsure	1	1.0	2.6	100.0
	Total	38	39.2	100.0	
Missing	Branching skip	51	52.6		
	Partial breakoff	5	5.2		
	No answer	3	3.1		
	Total	59	60.8		
Total		97	100.0		

Q08C How did the [Department] use the native seed that was wild collected on state land between 2017 and 2019? The seed was used for another purpose. Please specify.

		Frequency	Percent	Valid Percent	Cumulative Percent
Valid	Yes	7	7.2	24.1	24.1
	No	18	18.6	62.1	86.2
	Unsure	4	4.1	13.8	100.0
	Total	29	29.9	100.0	
Missing	Branching skip	51	52.6		
	Partial breakoff	5	5.2		
	No answer	12	12.4		
	Total	68	70.1		
Total		97	100.0		

Q09A In obtaining seed for [DEPARTMENT] projects between 2017-2019, how important was it to obtain this type of seed? Native seed

		Frequency	Percent	Valid Percent	Cumulative Percent
Valid	Very important	72	74.2	82.8	82.8
	Somewhat important	14	14.4	16.1	98.9
	Not important	1	1.0	1.1	100.0
	Total	87	89.7	100.0	
Missing	Branching skip	3	3.1		
	Partial breakoff	5	5.2		
	No answer	2	2.1		
	Total	10	10.3		
Total		97	100.0		

Q09B In obtaining seed for [DEPARTMENT] projects between 2017-2019, how important was it to obtain this type of seed? Certified native seed

		Frequency	Percent	Valid Percent	Cumulative Percent
Valid	Very important	37	38.1	42.5	42.5
	Somewhat important	28	28.9	32.2	74.7
	Not important	10	10.3	11.5	86.2
	Unsure	12	12.4	13.8	100.0
	Total	87	89.7	100.0	
Missing	Branching skip	3	3.1		
	Partial breakoff	5	5.2		
	No answer	2	2.1		
	Total	10	10.3		
Total		97	100.0		

Q09C In obtaining seed for [DEPARTMENT] projects between 2017-2019, how important was it to obtain this type of seed? Native seed that was collected from a designated geographic area or specified seed zone.

		Frequency	Percent	Valid Percent	Cumulative Percent
Valid	Very important	43	44.3	49.4	49.4
	Somewhat important	29	29.9	33.3	82.8
	Not important	10	10.3	11.5	94.3
	Unsure	5	5.2	5.7	100.0
	Total	87	89.7	100.0	
Missing	Branching skip	3	3.1		
	Partial breakoff	5	5.2		
	No answer	2	2.1		
	Total	10	10.3		
Total		97	100.0		

Q10 Were there any other important characteristics for the seed that was obtained during the 3-year period between 2017-2019?

		Frequency	Percent	Valid Percent	Cumulative Percent
Valid	Yes, please specify:	36	37.1	41.4	41.4
	No	51	52.6	58.6	100.0
	Total	87	89.7	100.0	
Missing	Branching skip	3	3.1		
	Partial breakoff	5	5.2		
	No answer	2	2.1		
	Total	10	10.3		
Total		97	100.0		

Q11A Were native seed or plant materials used for this purpose between 2017-2019? Pollinator.

		Frequency	Percent	Valid Percent	Cumulative Percent
Valid	Used	72	74.2	82.8	82.8
	Not used	10	10.3	11.5	94.3
	Unsure	5	5.2	5.7	100.0
	Total	87	89.7	100.0	
Missing	Branching skip	3	3.1		
	Partial breakoff	5	5.2		
	No answer	2	2.1		
	Total	10	10.3		
Total		97	100.0		

Q11B Were native seed or plant materials used for this purpose between 2017-2019? Creation/restoration of wildlife habitat (other than pollinator habitat projects).

		Frequency	Percent	Valid Percent	Cumulative Percent
Valid	Used	76	78.4	87.4	87.4
	Not used	6	6.2	6.9	94.3
	Unsure	5	5.2	5.7	100.0
	Total	87	89.7	100.0	
Missing	Branching skip	3	3.1		
	Partial breakoff	5	5.2		
	No answer	2	2.1		
	Total	10	10.3		
Total		97	100.0		

Q11C Were native seed or plant materials used for this purpose between 2017-2019? Restorative activity on land in a wilderness/natural area.

		Frequency	Percent	Valid Percent	Cumulative Percent
Valid	Used	65	67.0	74.7	74.7
	Not used	13	13.4	14.9	89.7
	Unsure	9	9.3	10.3	100.0
	Total	87	89.7	100.0	
Missing	Branching skip	3	3.1		
	Partial breakoff	5	5.2		
	No answer	2	2.1		
	Total	10	10.3		
Total		97	100.0		

Q11D Were native seed or plant materials used for this purpose between 2017-2019? Invasive species suppression.

		Frequency	Percent	Valid Percent	Cumulative Percent
Valid	Used	51	52.6	58.6	58.6
	Not used	24	24.7	27.6	86.2
	Unsure	12	12.4	13.8	100.0
	Total	87	89.7	100.0	
Missing	Branching skip	3	3.1		
	Partial breakoff	5	5.2		
	No answer	2	2.1		
	Total	10	10.3		
Total		97	100.0		

Q11E Were native seed or plant materials used for this purpose between 2017-2019? Natural disaster recovery from events such as a hurricane, flood, fire, severe drought, etc.

		Frequency	Percent	Valid Percent	Cumulative Percent
Valid	Used	32	33.0	37.2	37.2
	Not used	39	40.2	45.3	82.6
	Unsure	15	15.5	17.4	100.0
	Total	86	88.7	100.0	
Missing	Branching skip	3	3.1		
	Partial breakoff	5	5.2		
	No answer	3	3.1		
	Total	11	11.3		
Total		97	100.0		

Q11F Were native seed or plant materials used for this purpose between 2017-2019? Roadside seeding.

		Frequency	Percent	Valid Percent	Cumulative Percent
Valid	Used	58	59.8	66.7	66.7
	Not used	19	19.6	21.8	88.5
	Unsure	10	10.3	11.5	100.0
	Total	87	89.7	100.0	
Missing	Branching skip	3	3.1		
	Partial breakoff	5	5.2		
	No answer	2	2.1		
	Total	10	10.3		
Total		97	100.0		

Q11G Were native seed or plant materials used for this purpose between 2017-2019? Roadside maintenance.

		Frequency	Percent	Valid Percent	Cumulative Percent
Valid	Used	40	41.2	46.5	46.5
	Not used	27	27.8	31.4	77.9
	Unsure	19	19.6	22.1	100.0
	Total	86	88.7	100.0	
Missing	Branching skip	3	3.1		
	Partial breakoff	5	5.2		
	No answer	3	3.1		
	Total	11	11.3		
Total		97	100.0		

**Q11H Were native seed or plant materials used for this purpose between 2017-2019?
Stream erosion mitigation/restoration.**

		Frequency	Percent	Valid Percent	Cumulative Percent
Valid	Used	69	71.1	80.2	80.2
	Not used	10	10.3	11.6	91.9
	Unsure	7	7.2	8.1	100.0
	Total	86	88.7	100.0	
Missing	Branching skip	3	3.1		
	Partial breakoff	5	5.2		
	No answer	3	3.1		
	Total	11	11.3		
Total		97	100.0		

**Q11I Were native seed or plant materials used for this purpose
between 2017-2019? Soil protection.**

		Frequency	Percent	Valid Percent	Cumulative Percent
Valid	Used	63	64.9	72.4	72.4
	Not used	14	14.4	16.1	88.5
	Unsure	10	10.3	11.5	100.0
	Total	87	89.7	100.0	
Missing	Branching skip	3	3.1		
	Partial breakoff	5	5.2		
	No answer	2	2.1		
	Total	10	10.3		
Total		97	100.0		

**Q11J Were native seed or plant materials used for this purpose
between 2017-2019? Landscaping.**

		Frequency	Percent	Valid Percent	Cumulative Percent
Valid	Used	53	54.6	60.9	60.9
	Not used	24	24.7	27.6	88.5
	Unsure	10	10.3	11.5	100.0
	Total	87	89.7	100.0	
Missing	Branching skip	3	3.1		
	Partial breakoff	5	5.2		
	No answer	2	2.1		
	Total	10	10.3		
Total		97	100.0		

**Q11K Were native seed or plant materials used for this purpose
between 2017-2019? Green infrastructure.**

		Frequency	Percent	Valid Percent	Cumulative Percent
Valid	Used	43	44.3	49.4	49.4
	Not used	27	27.8	31.0	80.5
	Unsure	17	17.5	19.5	100.0
	Total	87	89.7	100.0	
Missing	Branching skip	3	3.1		
	Partial breakoff	5	5.2		
	No answer	2	2.1		
	Total	10	10.3		
Total		97	100.0		

**Q11L Were native seed or plant materials used for this purpose
between 2017-2019? Another purpose, please specify.**

		Frequency	Percent	Valid Percent	Cumulative Percent
Valid	Used	11	11.3	31.4	31.4
	Not used	9	9.3	25.7	57.1
	Unsure	15	15.5	42.9	100.0
	Total	35	36.1	100.0	
Missing	Branching skip	3	3.1		
	Partial breakoff	5	5.2		
	No answer	54	55.7		
	Total	62	63.9		
Total		97	100.0		

**Q12 Does [DEPARTMENT] have programs to assist private landowners with the use
of native seed or plant materials?**

		Frequency	Percent	Valid Percent	Cumulative Percent
Valid	Yes	40	41.2	46.0	46.0
	No	42	43.3	48.3	94.3
	Unsure	5	5.2	5.7	100.0
	Total	87	89.7	100.0	
Missing	Branching skip	3	3.1		
	Partial breakoff	5	5.2		
	No answer	2	2.1		
	Total	10	10.3		
Total		97	100.0		

Q13 Does [DEPARTMENT] contract out seeding or planting activities with native seed or plant materials?

		Frequency	Percent	Valid Percent	Cumulative Percent
Valid	Yes	68	70.1	78.2	78.2
	No	16	16.5	18.4	96.6
	Unsure	3	3.1	3.4	100.0
	Total	87	89.7	100.0	
Missing	Branching skip	3	3.1		
	Partial breakoff	5	5.2		
	No answer	2	2.1		
	Total	10	10.3		
Total		97	100.0		

Q14A Does the [DEPARTMENT] have this form of seed storage? Ambient storage.

		Frequency	Percent	Valid Percent	Cumulative Percent
Valid	Yes	56	57.7	64.4	64.4
	No	28	28.9	32.2	96.6
	Unsure	3	3.1	3.4	100.0
	Total	87	89.7	100.0	
Missing	Branching skip	3	3.1		
	Partial breakoff	5	5.2		
	No answer	2	2.1		
	Total	10	10.3		
Total		97	100.0		

Q14B Does the [DEPARTMENT] have this form of seed storage? Refrigerated storage.

		Frequency	Percent	Valid Percent	Cumulative Percent
Valid	Yes	24	24.7	27.6	27.6
	No	52	53.6	59.8	87.4
	Unsure	11	11.3	12.6	100.0
	Total	87	89.7	100.0	
Missing	Branching skip	3	3.1		
	Partial breakoff	5	5.2		
	No answer	2	2.1		
	Total	10	10.3		
Total		97	100.0		

Q14C Does the [DEPARTMENT] have this form of seed storage? Freezer storage.

		Frequency	Percent	Valid Percent	Cumulative Percent
Valid	Yes	14	14.4	16.3	16.3
	No	59	60.8	68.6	84.9
	Unsure	13	13.4	15.1	100.0
	Total	86	88.7	100.0	
Missing	Branching skip	3	3.1		
	Partial breakoff	5	5.2		
	No answer	3	3.1		
	Total	11	11.3		
Total		97	100.0		

Q15A Does [DEPARTMENT] use this type of contract? Marketing contracts that specifies the type, price, quantity, and delivery date of seed or plant materials.

		Frequency	Percent	Valid Percent	Cumulative Percent
Valid	Yes	32	33.0	38.6	38.6
	No	36	37.1	43.4	81.9
	Unsure	15	15.5	18.1	100.0
	Total	83	85.6	100.0	
Missing	Branching skip	7	7.2		
	Partial breakoff	5	5.2		
	No answer	2	2.1		
	Total	14	14.4		
Total		97	100.0		

Q15B Does [DEPARTMENT] use this type of contract? Production contracts that specifies the desired type, quantity and delivery date of native seed and plant materials.

		Frequency	Percent	Valid Percent	Cumulative Percent
Valid	Yes	18	18.6	22.0	22.0
	No	43	44.3	52.4	74.4
	Unsure	21	21.6	25.6	100.0
	Total	82	84.5	100.0	
Missing	Branching skip	7	7.2		
	Partial breakoff	5	5.2		
	No answer	3	3.1		
	Total	15	15.5		
Total		97	100.0		

Q16 Are the contracts usually established . . .

		Frequency	Percent	Valid Percent	Cumulative Percent
Valid	before the supplier begins the seed production process?	15	15.5	36.6	36.6
	after the supplier begins the seed production process?	10	10.3	24.4	61.0
	Unsure of timing of contract	16	16.5	39.0	100.0
	Total	41	42.3	100.0	
Missing	Branching skip	49	50.5		
	Partial breakoff	5	5.2		
	No answer	2	2.1		
	Total	56	57.7		
Total		97	100.0		

Q17A Do you use this source to obtain information about the availability of native seed and plant materials? In-house knowledge.

		Frequency	Percent	Valid Percent	Cumulative Percent
Valid	Used	86	88.7	97.7	97.7
	Not used	1	1.0	1.1	98.9
	Unsure	1	1.0	1.1	100.0
	Total	88	90.7	100.0	
Missing	Branching skip	3	3.1		
	Partial breakoff	5	5.2		
	No answer	1	1.0		
	Total	9	9.3		
Total		97	100.0		

Q17B Do you use this source to obtain information about the availability of native seed and plant materials? Advertising.

		Frequency	Percent	Valid Percent	Cumulative Percent
Valid	Used	33	34.0	40.7	40.7
	Not used	38	39.2	46.9	87.7
	Unsure	10	10.3	12.3	100.0
	Total	81	83.5	100.0	
Missing	Branching skip	3	3.1		
	Partial breakoff	5	5.2		
	No answer	8	8.2		
	Total	16	16.5		
Total		97	100.0		

Q17C Do you use this source to obtain information about the availability of native seed and plant materials? Preapproved vendors.

		Frequency	Percent	Valid Percent	Cumulative Percent
Valid	Used	60	61.9	69.8	69.8
	Not used	17	17.5	19.8	89.5
	Unsure	9	9.3	10.5	100.0
	Total	86	88.7	100.0	
Missing	Branching skip	3	3.1		
	Partial breakoff	5	5.2		
	No answer	3	3.1		
	Total	11	11.3		
Total		97	100.0		

Q17D Do you use this source to obtain information about the availability of native seed and plant materials? Request for proposals.

		Frequency	Percent	Valid Percent	Cumulative Percent
Valid	Used	43	44.3	51.2	51.2
	Not used	28	28.9	33.3	84.5
	Unsure	13	13.4	15.5	100.0
	Total	84	86.6	100.0	
Missing	Branching skip	3	3.1		
	Partial breakoff	5	5.2		
	No answer	5	5.2		
	Total	13	13.4		
Total		97	100.0		

Q17E Do you use this source to obtain information about the availability of native seed and plant materials? Other, please specify.

		Frequency	Percent	Valid Percent	Cumulative Percent
Valid	Used	19	19.6	51.4	51.4
	Not used	10	10.3	27.0	78.4
	Unsure	8	8.2	21.6	100.0
	Total	37	38.1	100.0	
Missing	Branching skip	3	3.1		
	Partial breakoff	5	5.2		
	No answer	52	53.6		
	Total	60	61.9		
Total		97	100.0		

Q18A Do you use this method to communicate to suppliers about your anticipated future need for native seed or plant materials? Requests for proposals.

		Frequency	Percent	Valid Percent	Cumulative Percent
Valid	Used	43	44.3	48.9	48.9
	Not used	37	38.1	42.0	90.9
	Unsure	8	8.2	9.1	100.0
	Total	88	90.7	100.0	
Missing	Branching skip	3	3.1		
	Partial breakoff	5	5.2		
	No answer	1	1.0		
	Total	9	9.3		
Total		97	100.0		

Q18B Do you use this method to communicate to suppliers about your anticipated future need for native seed or plant materials? General publicity around projects.

		Frequency	Percent	Valid Percent	Cumulative Percent
Valid	Used	26	26.8	29.9	29.9
	Not used	44	45.4	50.6	80.5
	Unsure	17	17.5	19.5	100.0
	Total	87	89.7	100.0	
Missing	Branching skip	3	3.1		
	Partial breakoff	5	5.2		
	No answer	2	2.1		
	Total	10	10.3		
Total		97	100.0		

Q18C Do you use this method to communicate to suppliers about your anticipated future need for native seed or plant materials? Conferences or other professional meetings.

		Frequency	Percent	Valid Percent	Cumulative Percent
Valid	Used	43	44.3	49.4	49.4
	Not used	36	37.1	41.4	90.8
	Unsure	8	8.2	9.2	100.0
	Total	87	89.7	100.0	
Missing	Branching skip	3	3.1		
	Partial breakoff	5	5.2		
	No answer	2	2.1		
	Total	10	10.3		
Total		97	100.0		

Q18D Do you use this method to communicate to suppliers about your anticipated future need for native seed or plant materials? Informal meetings with growers.

		Frequency	Percent	Valid Percent	Cumulative Percent
Valid	Used	39	40.2	45.3	45.3
	Not used	38	39.2	44.2	89.5
	Unsure	9	9.3	10.5	100.0
	Total	86	88.7	100.0	
Missing	Branching skip	3	3.1		
	Partial breakoff	5	5.2		
	No answer	3	3.1		
	Total	11	11.3		
Total		97	100.0		

Q18E Do you use this method to communicate to suppliers about your anticipated future need for native seed or plant materials? Word of mouth.

		Frequency	Percent	Valid Percent	Cumulative Percent
Valid	Used	48	49.5	54.5	54.5
	Not used	29	29.9	33.0	87.5
	Unsure	11	11.3	12.5	100.0
	Total	88	90.7	100.0	
Missing	Branching skip	3	3.1		
	Partial breakoff	5	5.2		
	No answer	1	1.0		
	Total	9	9.3		
Total		97	100.0		

Q18F Do you use this method to communicate to suppliers about your anticipated future need for native seed or plant materials? Other, please specify.

		Frequency	Percent	Valid Percent	Cumulative Percent
Valid	Used	9	9.3	28.1	28.1
	Not used	14	14.4	43.8	71.9
	Unsure	9	9.3	28.1	100.0
	Total	32	33.0	100.0	
Missing	Branching skip	3	3.1		
	Partial breakoff	5	5.2		
	No answer	57	58.8		
	Total	65	67.0		
Total		97	100.0		

Q19A Does this specification or guideline apply to [DEPARTMENT] projects that involve native seed or plant materials? Technical specifications.

		Frequency	Percent	Valid Percent	Cumulative Percent
Valid	Apply	79	81.4	90.8	90.8
	Do not apply	5	5.2	5.7	96.6
	Unsure	3	3.1	3.4	100.0
	Total	87	89.7	100.0	
Missing	Branching skip	3	3.1		
	Partial breakoff	5	5.2		
	No answer	2	2.1		
	Total	10	10.3		
Total		97	100.0		

Q19B Does this specification or guideline apply to [DEPARTMENT] projects that involve native seed or plant materials? Federal regulations, guidelines, or policy.

		Frequency	Percent	Valid Percent	Cumulative Percent
Valid	Apply	61	62.9	70.1	70.1
	Do not apply	12	12.4	13.8	83.9
	Unsure	14	14.4	16.1	100.0
	Total	87	89.7	100.0	
Missing	Branching skip	3	3.1		
	Partial breakoff	5	5.2		
	No answer	2	2.1		
	Total	10	10.3		
Total		97	100.0		

Q19C Does this specification or guideline apply to [DEPARTMENT] projects that involve native seed or plant materials? State regulations, guidelines, or policy.

		Frequency	Percent	Valid Percent	Cumulative Percent
Valid	Apply	73	75.3	83.9	83.9
	Do not apply	8	8.2	9.2	93.1
	Unsure	6	6.2	6.9	100.0
	Total	87	89.7	100.0	
Missing	Branching skip	3	3.1		
	Partial breakoff	5	5.2		
	No answer	2	2.1		
	Total	10	10.3		
Total		97	100.0		

Q19D Does this specification or guideline apply to [DEPARTMENT] projects that involve native seed or plant materials? Funding source specifications.

		Frequency	Percent	Valid Percent	Cumulative Percent
Valid	Apply	51	52.6	58.6	58.6
	Do not apply	17	17.5	19.5	78.2
	Unsure	19	19.6	21.8	100.0
	Total	87	89.7	100.0	
Missing	Branching skip	3	3.1		
	Partial breakoff	5	5.2		
	No answer	2	2.1		
	Total	10	10.3		
Total		97	100.0		

Q20 How often does [DEPARTMENT] substitute non-native seed or plant materials when native seed or plants are unavailable?

		Frequency	Percent	Valid Percent	Cumulative Percent
Valid	Frequently	21	21.6	24.1	24.1
	Infrequently	45	46.4	51.7	75.9
	Never	21	21.6	24.1	100.0
	Total	87	89.7	100.0	
Missing	Branching skip	3	3.1		
	Partial breakoff	5	5.2		
	No answer	2	2.1		
	Total	10	10.3		
Total		97	100.0		

Q21A Was this a reason substitutions with non-native seed or plant materials were typically made? Because native seed or plant material were available but too expensive.

		Frequency	Percent	Valid Percent	Cumulative Percent
Valid	Yes	28	28.9	43.1	43.1
	No	28	28.9	43.1	86.2
	Unsure	9	9.3	13.8	100.0
	Total	65	67.0	100.0	
Missing	Branching skip	24	24.7		
	Partial breakoff	5	5.2		
	No answer	3	3.1		
	Total	32	33.0		
Total		97	100.0		

Q21B Was this a reason substitutions with non-native seed or plant materials were typically made? Because native seed or plant material were available but not within the timeline needed.

		Frequency	Percent	Valid Percent	Cumulative Percent
Valid	Yes	22	22.7	33.8	33.8
	No	27	27.8	41.5	75.4
	Unsure	16	16.5	24.6	100.0
	Total	65	67.0	100.0	
Missing	Branching skip	24	24.7		
	Partial breakoff	5	5.2		
	No answer	3	3.1		
	Total	32	33.0		
Total		97	100.0		

Q21C Was this a reason substitutions with non-native seed or plant materials were typically made? Because native seed or plant material were unavailable regardless of price and timing.

		Frequency	Percent	Valid Percent	Cumulative Percent
Valid	Yes	39	40.2	58.2	58.2
	No	17	17.5	25.4	83.6
	Unsure	11	11.3	16.4	100.0
	Total	67	69.1	100.0	
Missing	Branching skip	24	24.7		
	Partial breakoff	5	5.2		
	No answer	1	1.0		
	Total	30	30.9		
Total		97	100.0		

Q21D Was this a reason substitutions with non-native seed or plant materials were typically made? Other reasons for the substitutions. Please specify.

		Frequency	Percent	Valid Percent	Cumulative Percent
Valid	Yes	24	24.7	70.6	70.6
	No	5	5.2	14.7	85.3
	Unsure	5	5.2	14.7	100.0
	Total	34	35.1	100.0	
Missing	Branching skip	24	24.7		
	Partial breakoff	5	5.2		
	No answer	34	35.1		
	Total	63	64.9		
Total		97	100.0		

Q22A How often does your agency substitute native seed or plant materials having different characteristics when your preferred natives are not available? Native seed or plant material of different species.

		Frequency	Percent	Valid Percent	Cumulative Percent
Valid	Frequently	36	37.1	41.9	41.9
	Infrequently	44	45.4	51.2	93.0
	Never	6	6.2	7.0	100.0
	Total	86	88.7	100.0	
Missing	Branching skip	3	3.1		
	Partial breakoff	5	5.2		
	No answer	3	3.1		
	Total	11	11.3		
Total		97	100.0		

Q22B How often does your agency substitute native seed or plant materials having different characteristics when your preferred natives are not available? Native seed or plant material from a different region.

		Frequency	Percent	Valid Percent	Cumulative Percent
Valid	Frequently	25	25.8	29.1	29.1
	Infrequently	51	52.6	59.3	88.4
	Never	10	10.3	11.6	100.0
	Total	86	88.7	100.0	
Missing	Branching skip	3	3.1		
	Partial breakoff	5	5.2		
	No answer	3	3.1		
	Total	11	11.3		
Total		97	100.0		

Q22C How often does your agency substitute native seed or plant materials having different characteristics when your preferred natives are not available? Other, please specify.

		Frequency	Percent	Valid Percent	Cumulative Percent
Valid	Frequently	2	2.1	10.5	10.5
	Infrequently	5	5.2	26.3	36.8
	Never	12	12.4	63.2	100.0
	Total	19	19.6	100.0	
Missing	Branching skip	6	6.2		
	Partial breakoff	5	5.2		
	No answer	67	69.1		
	Total	78	80.4		
Total		97	100.0		

Q23A Was this a reason substitutions with native seed or plant materials having different characteristics were typically made? Because native seed or plant material with the preferred characteristics were available but too expensive.

		Frequency	Percent	Valid Percent	Cumulative Percent
Valid	Yes	29	29.9	35.4	35.4
	No	39	40.2	47.6	82.9
	Unsure	14	14.4	17.1	100.0
	Total	82	84.5	100.0	
Missing	Branching skip	8	8.2		
	Partial breakoff	5	5.2		
	No answer	2	2.1		
	Total	15	15.5		
Total		97	100.0		

Q23B Was this a reason substitutions with native seed or plant materials having different characteristics were typically made? Because native seed or plant material with the preferred characteristics were available but not within the timeline needed.

		Frequency	Percent	Valid Percent	Cumulative Percent
Valid	Yes	49	50.5	60.5	60.5
	No	22	22.7	27.2	87.7
	Unsure	10	10.3	12.3	100.0
	Total	81	83.5	100.0	
Missing	Branching skip	8	8.2		
	Partial breakoff	5	5.2		
	No answer	3	3.1		
	Total	16	16.5		
Total		97	100.0		

Q23C Was this a reason substitutions with native seed or plant materials having different characteristics were typically made? Because native seed or plant material with the preferred characteristics were unavailable regardless of price and timing.

		Frequency	Percent	Valid Percent	Cumulative Percent
Valid	Yes	68	70.1	81.0	81.0
	No	8	8.2	9.5	90.5
	Unsure	8	8.2	9.5	100.0
	Total	84	86.6	100.0	
Missing	Branching skip	8	8.2		
	Partial breakoff	5	5.2		
	Total	13	13.4		
Total		97	100.0		

Q23D Was this a reason substitutions with native seed or plant materials having different characteristics were typically made? Other reasons for the substitutions. Please specify.

		Frequency	Percent	Valid Percent	Cumulative Percent
Valid	Yes	8	8.2	32.0	32.0
	No	10	10.3	40.0	72.0
	Unsure	7	7.2	28.0	100.0
	Total	25	25.8	100.0	
Missing	Branching skip	8	8.2		
	Partial breakoff	5	5.2		
	No answer	59	60.8		
	Total	72	74.2		
Total		97	100.0		

Q24 In order to evaluate [DEPARTMENT] projects that involve native seed or plant materials, do you check on the survival of the seed or plant materials after planting?

		Frequency	Percent	Valid Percent	Cumulative Percent
Valid	Yes, for most project	63	64.9	71.6	71.6
	Yes, for some projects	23	23.7	26.1	97.7
	No	2	2.1	2.3	100.0
	Total	88	90.7	100.0	
Missing	Branching skip	3	3.1		
	Partial breakoff	5	5.2		
	No answer	1	1.0		
	Total	9	9.3		
Total		97	100.0		

Q25 For a typical project, when is the last time you checked on the survival of the seed or plant materials after planting?

		Frequency	Percent	Valid Percent	Cumulative Percent
Valid	About 1 year or less after seeding/planting	21	21.6	24.4	24.4
	1 to 3 years after seeding/planting	39	40.2	45.3	69.8
	Greater than 3 years after seeding/planting	26	26.8	30.2	100.0
	Total	86	88.7	100.0	
Missing	Branching skip	5	5.2		
	Partial breakoff	5	5.2		
	No answer	1	1.0		
	Total	11	11.3		
Total		97	100.0		

Q26 Does [DEPARTMENT] have the range of in-house expertise needed for the various aspects of projects that use native seeds or plants, or are there some gaps in the expertise available in-house?

		Frequency	Percent	Valid Percent	Cumulative Percent
Valid	We have the expertise we need in-house	51	52.6	58.0	58.0
	We have some gaps in expertise. Pease specify:	34	35.1	38.6	96.6
	Don"t know	3	3.1	3.4	100.0
	Total	88	90.7	100.0	
Missing	Branching skip	3	3.1		
	Partial breakoff	5	5.2		
	No answer	1	1.0		
	Total	9	9.3		
Total		97	100.0		

Q27_O What are the biggest barriers or disincentives to using native seed and plant materials in your work? (Open-ended response coding frequencies)

Code	Count
1. Availability	48
2. Cost	38
3. Communication issues	1
4. Genetics	16
5. Geographic or ecosystem limitations conditions	9
7. Systems	5
8. Timing	9
9 Other	0
10. Lack of knowledge	11
11. Lack of support for native plants/seeds	13
11. Not applicable, no comment	7

Q28 Do you anticipate that the use of native seed or plant materials by the [DEPARTMENT] is likely to increase or decrease in the near-term future?

		Frequency	Percent	Valid Percent	Cumulative Percent
Valid	Increase	65	67.0	75.6	75.6
	Decrease	2	2.1	2.3	77.9
	Unsure	19	19.6	22.1	100.0
	Total	86	88.7	100.0	
Missing	Branching skip	3	3.1		
	Partial breakoff	5	5.2		
	No answer	3	3.1		
	Total	11	11.3		
Total		97	100.0		

Q28_O Do you anticipate that the use of native seed or plant materials by the [DEPARTMENT] is likely to increase or decrease in the near-term future? Please explain why. (Open-ended response coding frequencies)

Increase	
CODE	**Count**
Preference, demand, interest	13
Research, Funding	6
Relationships	4
Environmental reasons	8
Awareness, education	11
Habitat programs, restoration	12
Projects	7
Decrease	
CODE	
Staffing	1
Unsure	
CODE	
Funding dependent	3
Natives already	4
Climate dependent	1
Innovation dependent	1
Remain the same	3

Q29 Do you anticipate that [DEPARTMENT]'s use of native seed or plant materials is likely to increase or decrease in the long-term future?

		Frequency	Percent	Valid Percent	Cumulative Percent
Valid	Increase	62	63.9	72.9	72.9
	Decrease	1	1.0	1.2	74.1
	Unsure	22	22.7	25.9	100.0
	Total	85	87.6	100.0	
Missing	Branching skip	3	3.1		
	Partial breakoff	5	5.2		
	No answer	4	4.1		
	Total	12	12.4		
Total		97	100.0		

Q29_O Do you anticipate that [DEPARTMENT]'s use of native seed or plant materials is likely to increase or decrease in the long-term future? Please explain why. (Open-ended response coding frequencies)

Increase	
CODE	**Count**
Awareness, education	7
Preference, demand, interest	13
Environmental reasons	7
Funding, research	4
Population growth	2
Habitat programs, restoration	9
See Q28	5
Unsure	
CODE	
Funding, research dependent	1
Markets	3
Project dependent	2
Needs dependent	2
Climate dependent	2
See Q28	3
No answer	
Natives already	1
See Q28	1

Q30 Thinking about the three-year period from 2017 to 2019, what was [DEPARTMENT]'s approximate average annual expenditure on native seed and plant materials combined? (Open-ended response coding frequencies)

	Total	
	Freq	%
$100,000 and below	43	56%
Over $100,000	26	34%
Don't know	8	10%
Total	77	100%
Missing	20	
Total	97	

Q31_O If you have any additional comments, please feel free to type them in the space provided below. Thank you for your help! (Open-ended response coding frequencies)

DESCRIPTION	count
Talked about managing or using nonnative seeds/plants and/or why non-native seed/plants are used instead of native seed/plants.	4
Talked about research or need for research.	4
Questioning whether the term "native" to a region is actually true, and/or difficult to determine the origin of non-certified seed	2
Talked about something specific to a survey question or how they answered.	8
Specific to what their business does	11
Said NA	1

Appendix 2E

Supplier Survey Invitation Letter

The National
Academies of

SCIENCES
ENGINEERING
MEDICINE

May 2021

\<Business Name\>
\<Business Address\>
\<City\>, \<State\> \<Zip\>-\<Zip4\>

The National Academies of Sciences, Engineering, and Medicine is conducting an important assessment of the challenges faced by seed and plant suppliers in the United States. I am writing because it is my understanding that your organization is a supplier of native seed or plant materials. The assessment will benefit suppliers throughout the nation by identifying ways for improving connections between supply and demand for native seed and plant materials.

We would greatly appreciate your asking the person at [Business Name] who is <u>most knowledgeable</u> about your sales of native seed and plant materials to answer this brief online survey that is being sent to seed and plant suppliers throughout the United States. The survey asks about experiences your organization might have with the sale of native seed. We would also like to hear your ideas for improving future opportunities for native seed and plant suppliers.

The online survey is located here:
https://opinion.wsu.edu/SeedSales Your survey
access code is: **\<RespID\>**

Responses to this survey are voluntary and your answers will be kept completely confidential by the National Academies and the Washington State University Social and Economic Sciences Research Center that is implementing the survey for us. No individual or business will ever be identified in our results. For this study to be successful it is vitally important that we hear from as many suppliers as possible.

If you have any questions about why this survey is being done, please contact Krisztina Marton at the National Academies at KMarton@nas.edu. If you have questions about accessing the survey, please contact Thom Allen at Washington State University, ted@wsu.edu, or call him toll-free at 1-800-833-0867.

Thank you for considering our request and we hope to hear from you soon.

Sincerely,

Robin Schoen
Director, Board on Agriculture and Natural Resources
The National Academies of Sciences, Engineering, and Medicine

Appendix 2F

Supplier Web Survey Instrument

Throughout this survey, by "<u>native</u>" we mean seed or plants indigenous to North America prior to European settlement. "<u>Plant materials</u>" could include, for example, plants, seedlings, and vegetative materials (other than seed).

First, please indicate which of the following does your business sell and does not sell. *Please mark one answer in each row.*

	Sell	Does not sell
Native seed	○	○
Non-native seed	○	○
Native plant materials, other than seed	○	○
Non-native plant materials, other than seed	○	○

<< Back | Next >>

[BRANCH: IF "DOES NOT SELL' TO BOTH NATIVE SEED AND NATIVE PLANT MATERIALS, END SURVEY]

Are the seed and plant material sales at your business mainly focused on selling to the home market, such as the sale of seed packets or individual plants for residential projects?

○ Yes

○ No

<< Back | Next >>

[BRANCH: IF "Yes" TO Q2 (Focus on home market), END SURVEY] [BRANCH: IF "No" TO Q1a (Native seed), GO TO Q6]

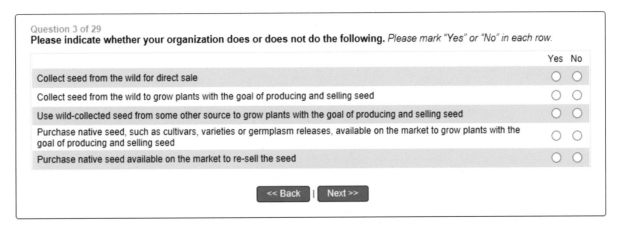

Please indicate whether your organization does or does not do the following. *Please mark "Yes" or "No" in each row.*

	Yes	No
Collect seed from the wild for direct sale	○	○
Collect seed from the wild to grow plants with the goal of producing and selling seed	○	○
Use wild-collected seed from some other source to grow plants with the goal of producing and selling seed	○	○
Purchase native seed, such as cultivars, varieties or germplasm releases, available on the market to grow plants with the goal of producing and selling seed	○	○
Purchase native seed available on the market to re-sell the seed	○	○

<< Back | Next >>

Question 4 of 29

For each of the following, please indicate whether you sell or do not sell this type of seed. *Please mark one answer in each row.*

	Sell	Do not sell	Unsure
Uncertified seed of claimed provenance of wild origin	○	○	○
Certified source-identified seed	○	○	○
Native germplasm releases or cultivars	○	○	○
Seed mixes that include native seed and comply with USDA conservation program requirements, such as the Conservation Reserve Program	○	○	○
Seed types that meet the requirements of the Bureau of Land Management	○	○	○

<< Back | Next >>

Question 5 of 29

What percentage of your organization's total native seed sales do each of the categories below represent? *Your best estimate is fine.*

	Percent of native seed sales
Native grass seed	[ ] %
Native forbs seed	[ ] %
Native shrubs seed	[ ] %
Native tree seed	[ ] %
Other native seed	[ ] %
Total	100 %

☐ Check this box if you are unable to provide this information.

<< Back | Next >>

[BRANCH: IF 'NO' TO Q1c (Native plant materials) SKIP TO Q8.]

Question 6 of 29
Does your organization . . . *Please mark one answer in each row.*

	Yes	No
Grow native plants to sell?	○	○
Purchase and re-sell native plant materials available on the market?	○	○

<< Back | Next >>

Question 7 of 29
What percentage of your organization's total native plant material sales do each of the categories below represent? *Your best estimate is fine.*

	Percent of native plant materials sales	
Native grasses		%
Native forbs		%
Native shrubs		%
Native trees		%
Other native plant materials		%
Total	100	%

☐ Check this box if you are unable to provide this information.

<< Back | Next >>

Question 8 of 29
Do you sell native plants or seeds sourced from different locations to match different geographical conditions or seed zones, as in different ecotypes of the same species?

○ Yes
○ No

<< Back | Next >>

Question 9 of 29

Thinking about an average over the three-year period from 2017 to 2019, for each of these buyer types, please indicate whether they represented a major portion, minor portion, or none of your annual sales of native seed and plant materials. *Please mark one answer in each row.*

	Major portion of sales	Minor portion of sales	None of the sales
Federal government agencies	○	○	○
State government agencies	○	○	○
Municipalities	○	○	○
Tribal governments	○	○	○
Private contractors	○	○	○
Non-profit organizations	○	○	○
Farmers and ranchers	○	○	○

<< Back | Next >>

Question 10 of 29

Do you sell services? *By services, we mean, for example, seeding operations, planting operations, seed cleaning, or other similar activities.*

○ Yes. Which services? Please specify:

○ No

<< Back | Next >>

Question 11 of 29

For each of the following types of storage, does the availability of this type of storage, either in terms of amount or cost, limit the quantity of seed you can sell? *Please mark "Yes" or "No" in each row.*

	Limits it	Does not limit it
Ambient storage	○	○
Refrigerated storage	○	○
Freezer storage	○	○

<< Back | Next >>

Question 12 of 29

For each of the following, please indicate whether you use or do not use this method to communicate to potential buyers about what your business offers. *Please mark one answer in each row.*

	Use	Do not use
Company website	○	○
Print catalogue	○	○
Vendor listings or registries	○	○
Bids in response to "requests for proposals" or "seed buy requests"	○	○
Word-of-mouth	○	○
Other, please specify: _____	○	○

<< Back | Next >>

Question 13 of 29

For each of the following contracting arrangements, please indicate whether you use them with buyers. *Please mark "Yes" or "No" in each row.*

	Yes	No
Bids on periodic consolidated seed buys, such as the Bureau of Land Management or state seed buys	○	○
Other spot market sales of available seed or plant materials. Click here for a definition.	○	○
Marketing contracts that specify the type, price, quantity, and delivery date of seed or plant materials. Click here for a definition.	○	○
Production contracts that specify the desired type, quantity and delivery date of seed or plant materials. It may or may not specify a price. Click here for a definition.	○	○

<< Back | Next >>

Question 14 of 29

Generally speaking, how important or not important are each of these characteristics to you when considering a contract? *Please mark one answer in each row.*

	Very important	Somewhat important	Not important	No opinion
Price guarantee	○	○	○	○
Purchase guarantee for predetermined quantity	○	○	○	○
Production cost sharing	○	○	○	○
Multi-production cycle contract	○	○	○	○
Delivery timeline	○	○	○	○
Prior positive experience with the buyer	○	○	○	○

<< Back | Next >>

Question 15 of 29
Relative to when you would start to produce native seeds or plant materials for sale, when do you normally sign a contract to sell these seeds or plant materials? *Choose the answer that is most typical for your business.*

○ Prior to when you would collect or purchase foundation seed

○ Prior to when you would begin multiplying from foundation seed in your possession

○ Prior to when you would begin producing the seeds or plant materials for sale, in other words, forward contract before production begins

○ After seed production begins but before harvest of all seed or plant material to be sold, in other words, forward contract after production is underway

○ After seed production is complete, in other words, sale from inventory

<< Back | Next >>

Question 16 of 29
For each of the following types of information, please indicate what role it plays in how you anticipate future demand for native seed or plant materials in order to plan ahead. *Please mark one answer in each row.*

	Major role	Moderate role	Minor role	No role
Bureau of Land Management (BLM) seed buy history	○	○	○	○
Past purchases or requests from other buyers	○	○	○	○
Information about forward contracting opportunities	○	○	○	○
Conservation Research Program funding and plans	○	○	○	○
Wildfire activity	○	○	○	○
Oil and gas leases	○	○	○	○
Mining operations	○	○	○	○
Native seed use trends, such as the Monarch Highway	○	○	○	○
Demand for urban and home landscaping	○	○	○	○
Other, please describe:	○	○	○	○

<< Back | Next >>

Question 17 of 29
Thinking about the average over the three-year period from 2017 to 2019, approximately what percentage of your inventory of native seed and plant materials was left unsold at the end of the marketing year? *Your best estimate is fine.*

[   ] % Percent unsold

<< Back | Next >>

Question 18 of 29
Over the three-year period from 2017 to 2019, was the average percentage of native seed and plant materials that was left unsold at the end of the marketing year . . .

○ Less than you anticipated

○ About what you anticipated

○ More than you anticipated

<< Back | Next >>

Question 19 of 29
For each of the following, please indicate the extent to which they are a challenge or not a challenge for you as a supplier of native seed or plant materials. *Please mark one answer in each row.*

	Major challenge	Moderate challenge	Minor challenge	Not a challenge
Insufficient demand	○	○	○	○
Unpredictable demand	○	○	○	○
Demand for plants without propagation protocols	○	○	○	○
Variability in seed testing results	○	○	○	○
Species difficult to grow	○	○	○	○
Lack of stock seed from appropriate seed zones or other specified locations	○	○	○	○
Other, please describe:	○	○	○	○

<< Back | Next >>

[BRANCH: IF REPLIED "NOT A CHALLENGE" TO EVERYTHING, SKIP TO Q21]

Question 20 of 29
What suggestions do you have that might help address the challenges you have encountered as a supplier of native seeds or plant materials?

<< Back | Next >>

[BRANCH: If YES TO Q3a OR Q3b (COLLECTS NATIVE SEED), CONTINUE. OTHERWISE SKIP TO Q22

Question 21 of 29

If you were to anticipate higher demand for <u>directly wild-collected native seed</u>, would your business be able to expand to collect more seed?

○ Likely yes

○ Likely no

<< Back | Next >>

Question 22 of 29

To the extent you are aware, are there major barriers or disincentives to wild-collecting native seed?

○ Yes. What are those barriers? [_____]

○ No

<< Back | Next >>

[BRANCH: If YES TO Q3b, Q3c, or Q3d (GROWER OF NATIVE SEED) CONTINUE. OTHERWISE SKIP TO Q24]

Question 23 of 29

If you were to anticipate higher demand, would your business be able to expand <u>to grow more native plants with the goal of producing and selling native seed</u>?

○ Likely yes

○ Likely no

<< Back | Next >>

Question 24 of 29

To the extent you are aware, are there any major barriers or disincentives to growing native plants with the goal of producing and selling native seed?

○ Yes. What are those barriers? [_____]

○ No

<< Back | Next >>

[BRANCH: If YES TO Q6a (GROWS PLANT MATERIALS TO SELL) CONTINUE. OTHERWISE SKIP TO Q26]

Question 25 of 29
If you were to anticipate higher demand, would your business be able to expand to <u>grow and sell more native plant materials</u>?

○ Likely yes

○ Likely no

<< Back | Next >>

Question 26 of 29
To the extent you are aware, are there any major barriers or disincentives to growing native plant materials to sell?

○ Yes. What are those barriers? [_____]

○ No

<< Back | Next >>

Question 27 of 29
Do you have any other suggestions that could improve how the native seed market functions?

[_____]

<< Back | Next >>

Question 28 of 29
The final two questions ask about your average annual sales and operating revenues. Having a sense, in general terms, of the sizes of the seed suppliers completing this survey will help us better understand the challenges faced by vendors of different sizes.

Thinking about the three-year period from 2017 to 2019, which of the following best describes the average annual sales and operating revenues of your business? *Your best estimate is fine.*

○ Less than $100,000

○ $100,000-$499,999

○ $500,000-$999,999

○ $1,000,000-$4,999,999

○ $5,000,000-$10,000,000

○ Over $10,000,000

<< Back | Next >>

Question 29 of 29
Roughly what percentage of your average annual sales and operating revenues are from the sale of native seed and plant materials? *Your best estimate is fine.*

[ ] % Percent from native seed and plant materials

<< Back | Next >>

If you have any additional comments, please feel free to type them in the space provided below.

Thank you!

<< Back | Next >>

You are about to finish this questionnaire.

To submit the questionnaire click the "Submit questionnaire" button below.

To review your answers starting from the beginning click the "Review your answers" button.

Review your answers | Submit questionnaire

Your survey has been received.

Please close this window.

Appendix 2G

Supplier Survey Frequency Distributions

Q01A Does your business sell any of the following products? Native seed.

		Frequency	Percent	Valid Percent	Cumulative Percent
Valid	Sell	141	52.0	56.0	56.0
	Does not sell	111	41.0	44.0	100.0
	Total	252	93.0	100.0	
Missing	Dont know (phone only)	1	.4		
	No answer	18	6.6		
	Total	19	7.0		
Total		271	100.0		

Q01B Does your business sell any of the following products? Non-native seed.

		Frequency	Percent	Valid Percent	Cumulative Percent
Valid	Sell	77	28.4	33.0	33.0
	Does not sell	156	57.6	67.0	100.0
	Total	233	86.0	100.0	
Missing	Dont know (phone only)	3	1.1		
	Partial breakoff	1	.4		
	No answer	34	12.5		
	Total	38	14.0		
Total		271	100.0		

Q01C Does your business sell any of the following products? Native plant materials, other than seed.

		Frequency	Percent	Valid Percent	Cumulative Percent
Valid	Sell	202	74.5	81.1	81.1
	Does not sell	47	17.3	18.9	100.0
	Total	249	91.9	100.0	
Missing	Partial breakoff	2	.7		
	No answer	20	7.4		
	Total	22	8.1		
Total		271	100.0		

185

Q01D Does your business sell any of the following products? Non-native plant materials, other than seed.

		Frequency	Percent	Valid Percent	Cumulative Percent
Valid	Sell	102	37.6	44.5	44.5
	Does not sell	127	46.9	55.5	100.0
	Total	229	84.5	100.0	
Missing	Dont know (phone only)	2	.7		
	Partial breakoff	3	1.1		
	No answer	37	13.7		
	Total	42	15.5		
Total		271	100.0		

Q02 Are the seed and plant material sales at your business mainly focused on selling to the home market, such as the sale of seed packets or individual plants for residential projects?

		Frequency	Percent	Valid Percent	Cumulative Percent
Valid	No	249	91.9	100.0	100.0
Missing	Partial breakoff	12	4.4		
	No answer	10	3.7		
	Total	22	8.1		
Total		271	100.0		

Q03A Does your organization do this? Collect seed from the wild for direct sale.

		Frequency	Percent	Valid Percent	Cumulative Percent
Valid	Yes	81	29.9	55.5	55.5
	No	65	24.0	44.5	100.0
	Total	146	53.9	100.0	
Missing	Branching skip	100	36.9		
	Partial breakoff	21	7.7		
	No answer	4	1.5		
	Total	125	46.1		
Total		271	100.0		

Q03B Does your organization do this? Collect seed from the wild to grow plants with the goal of producing and selling seed.

		Frequency	Percent	Valid Percent	Cumulative Percent
Valid	Yes	80	29.5	56.7	56.7
	No	61	22.5	43.3	100.0
	Total	141	52.0	100.0	
Missing	Branching skip	100	36.9		
	Partial breakoff	21	7.7		
	No answer	9	3.3		
	Total	130	48.0		
Total		271	100.0		

Q03C Does your organization do this? Use wild-collected seed from some other source to grow plants with the goal of producing and selling seed.

		Frequency	Percent	Valid Percent	Cumulative Percent
Valid	Yes	66	24.4	46.8	46.8
	No	75	27.7	53.2	100.0
	Total	141	52.0	100.0	
Missing	Branching skip	100	36.9		
	Partial breakoff	21	7.7		
	No answer	9	3.3		
	Total	130	48.0		
Total		271	100.0		

Q03D Does your organization do this? Purchase native seed, such as cultivars, varieties or germplasm releases, available on the market to grow plants with the goal of producing and selling seed.

		Frequency	Percent	Valid Percent	Cumulative Percent
Valid	Yes	46	17.0	32.6	32.6
	No	95	35.1	67.4	100.0
	Total	141	52.0	100.0	
Missing	Branching skip	100	36.9		
	Partial breakoff	21	7.7		
	No answer	9	3.3		
	Total	130	48.0		
Total		271	100.0		

Q03E Does your organization do this? Purchase native seed available on the market to re-sell the seed.

		Frequency	Percent	Valid Percent	Cumulative Percent
Valid	Yes	94	34.7	64.4	64.4
	No	52	19.2	35.6	100.0
	Total	146	53.9	100.0	
Missing	Branching skip	100	36.9		
	Partial breakoff	21	7.7		
	No answer	4	1.5		
	Total	125	46.1		
Total		271	100.0		

Q04A Do you sell this type of seed? Uncertified seed of claimed provenance of wild origin.

		Frequency	Percent	Valid Percent	Cumulative Percent
Valid	Sell	97	35.8	67.8	67.8
	Does not sell	38	14.0	26.6	94.4
	Unsure	8	3.0	5.6	100.0
	Total	143	52.8	100.0	
Missing	Branching skip	100	36.9		
	Partial breakoff	23	8.5		
	No answer	5	1.8		
	Total	128	47.2		
Total		271	100.0		

Q04B Do you sell this type of seed? Certified source-identified seed.

		Frequency	Percent	Valid Percent	Cumulative Percent
Valid	Sell	92	33.9	64.8	64.8
	Does not sell	47	17.3	33.1	97.9
	Unsure	3	1.1	2.1	100.0
	Total	142	52.4	100.0	
Missing	Branching skip	100	36.9		
	Partial breakoff	23	8.5		
	No answer	6	2.2		
	Total	129	47.6		
Total		271	100.0		

Q04C Do you sell this type of seed? Native germplasm releases or cultivars.

		Frequency	Percent	Valid Percent	Cumulative Percent
Valid	Sell	56	20.7	39.7	39.7
	Does not sell	74	27.3	52.5	92.2
	Unsure	11	4.1	7.8	100.0
	Total	141	52.0	100.0	
Missing	Branching skip	100	36.9		
	Partial breakoff	23	8.5		
	No answer	7	2.6		
	Total	130	48.0		
Total		271	100.0		

Q04D Do you sell this type of seed? Seed mixes that include native seed and comply with USDA conservation program requirements.

		Frequency	Percent	Valid Percent	Cumulative Percent
Valid	Sell	64	23.6	44.4	44.4
	Does not sell	61	22.5	42.4	86.8
	Unsure	19	7.0	13.2	100.0
	Total	144	53.1	100.0	
Missing	Branching skip	100	36.9		
	Partial breakoff	23	8.5		
	No answer	4	1.5		
	Total	127	46.9		
Total		271	100.0		

Q04E Do you sell this type of seed? Seed types that meet the requirements of the Bureau of Land Management.

		Frequency	Percent	Valid Percent	Cumulative Percent
Valid	Sell	50	18.5	34.7	34.7
	Does not sell	46	17.0	31.9	66.7
	Unsure	48	17.7	33.3	100.0
	Total	144	53.1	100.0	
Missing	Branching skip	100	36.9		
	Partial breakoff	23	8.5		
	No answer	4	1.5		
	Total	127	46.9		
Total		271	100.0		

Q05a What percentage of your organization's total native seed sales does this represent? Native grass seed.

		Frequency	Percent	Valid Percent	Cumulative Percent
Valid	0	32	11.8	26.4	26.4
	5	5	1.8	4.1	30.6
	8	1	.4	.8	31.4
	10	8	3.0	6.6	38.0
	15	1	.4	.8	38.8
	20	6	2.2	5.0	43.8
	24	1	.4	.8	44.6
	25	3	1.1	2.5	47.1
	30	5	1.8	4.1	51.2
	37	1	.4	.8	52.1
	40	4	1.5	3.3	55.4
	45	4	1.5	3.3	58.7
	48	3	1.1	2.5	61.2
	49	2	.7	1.7	62.8
	50	7	2.6	5.8	68.6
	55	1	.4	.8	69.4
	60	6	2.2	5.0	74.4
	62	1	.4	.8	75.2
	70	8	3.0	6.6	81.8
	75	2	.7	1.7	83.5
	80	3	1.1	2.5	86.0
	85	6	2.2	5.0	90.9
	90	4	1.5	3.3	94.2
	97	1	.4	.8	95.0
	98	2	.7	1.7	96.7
	100	4	1.5	3.3	100.0
	Total	121	44.6	100.0	
Missing	Branching skip	24	8.9		
	Partial breakoff	27	10.0		
	No answer	99	36.5		
	Total	150	55.4		
Total		271	100.0		

Q05b What percentage of your organization's total native seed sales does this represent? Native forbs seed.

		Frequency	Percent	Valid Percent	Cumulative Percent
Valid	0	26	9.6	21.5	21.5
	1	3	1.1	2.5	24.0
	2	1	.4	.8	24.8
	4	1	.4	.8	25.6
	5	4	1.5	3.3	28.9
	9	1	.4	.8	29.8
	10	13	4.8	10.7	40.5
	15	4	1.5	3.3	43.8
	17	1	.4	.8	44.6
	19	1	.4	.8	45.5
	20	4	1.5	3.3	48.8
	24	1	.4	.8	49.6
	25	3	1.1	2.5	52.1
	30	7	2.6	5.8	57.9
	35	1	.4	.8	58.7
	40	5	1.8	4.1	62.8
	45	4	1.5	3.3	66.1
	48	1	.4	.8	66.9
	49	1	.4	.8	67.8
	50	11	4.1	9.1	76.9
	55	1	.4	.8	77.7
	58	1	.4	.8	78.5
	60	2	.7	1.7	80.2
	65	1	.4	.8	81.0
	70	5	1.8	4.1	85.1
	75	4	1.5	3.3	88.4
	80	4	1.5	3.3	91.7
	85	1	.4	.8	92.6
	90	4	1.5	3.3	95.9
	95	2	.7	1.7	97.5
	100	3	1.1	2.5	100.0
	Total	121	44.6	100.0	
Missing	Branching skip	24	8.9		
	Partial breakoff	27	10.0		
	No answer	99	36.5		
	Total	150	55.4		
Total		271	100.0		

Q05c What percentage of your organization's total native seed sales does this represent? Native shrubs seed.

		Frequency	Percent	Valid Percent	Cumulative Percent
Valid	0	44	16.2	41.9	41.9
	1	1	.4	1.0	42.9
	1	4	1.5	3.8	46.7
	2	6	2.2	5.7	52.4
	3	1	.4	1.0	53.3
	5	19	7.0	18.1	71.4
	6	2	.7	1.9	73.3
	8	1	.4	1.0	74.3
	10	5	1.8	4.8	79.0
	15	1	.4	1.0	80.0
	20	6	2.2	5.7	85.7
	24	1	.4	1.0	86.7
	25	4	1.5	3.8	90.5
	30	3	1.1	2.9	93.3
	35	1	.4	1.0	94.3
	45	1	.4	1.0	95.2
	50	3	1.1	2.9	98.1
	90	1	.4	1.0	99.0
	100	1	.4	1.0	100.0
	Total	105	38.7	100.0	
Missing	Branching skip	24	8.9		
	Partial breakoff	27	10.0		
	No answer	115	42.4		
	Total	166	61.3		
Total		271	100.0		

Q05d What percentage of your organization's total native seed sales does this represent? Native tree seed.

		Frequency	Percent	Valid Percent	Cumulative Percent
Valid	0	56	20.7	58.3	58.3
	1	5	1.8	5.2	63.5
	2	1	.4	1.0	64.6
	4	1	.4	1.0	65.6
	5	2	.7	2.1	67.7
	10	4	1.5	4.2	71.9
	15	2	.7	2.1	74.0
	20	3	1.1	3.1	77.1
	24	1	.4	1.0	78.1
	25	2	.7	2.1	80.2
	30	1	.4	1.0	81.3
	45	1	.4	1.0	82.3
	50	5	1.8	5.2	87.5
	55	1	.4	1.0	88.5
	65	1	.4	1.0	89.6
	70	1	.4	1.0	90.6
	98	1	.4	1.0	91.7
	100	8	3.0	8.3	100.0
	Total	96	35.4	100.0	
Missing	Branching skip	24	8.9		
	Partial breakoff	27	10.0		
	No answer	124	45.8		
	Total	175	64.6		
Total		271	100.0		

Q05e What percentage of your organization's total native seed sales does this represent? Other native seed.

		Frequency	Percent	Valid Percent	Cumulative Percent
Valid	0	53	19.6	60.2	60.2
	1	3	1.1	3.4	63.6
	2	4	1.5	4.5	68.2
	3	1	.4	1.1	69.3
	4	3	1.1	3.4	72.7
	5	11	4.1	12.5	85.2
	10	3	1.1	3.4	88.6
	11	1	.4	1.1	89.8
	20	2	.7	2.3	92.0
	25	1	.4	1.1	93.2
	29	1	.4	1.1	94.3
	40	3	1.1	3.4	97.7
	50	2	.7	2.3	100.0
	Total	88	32.5	100.0	
Missing	Branching skip	24	8.9		
	Partial breakoff	27	10.0		
	No answer	132	48.7		
	Total	183	67.5		
Total		271	100.0		

Q05a_dk What percentage of your organization's total native seed sales does this represent? Check this box if you are unable to provide this information.

		Frequency	Percent	Valid Percent	Cumulative Percent
Valid	Checked	9	3.3	5.5	5.5
	Not checked	156	57.6	94.5	100.0
	Total	165	60.9	100.0	
Missing	Branching skip	100	36.9		
	Partial breakoff	6	2.2		
	Total	106	39.1		
Total		271	100.0		

Q06A Does your organization . . . Grow native plants to sell.

		Frequency	Percent	Valid Percent	Cumulative Percent
Valid	Yes	165	60.9	85.5	85.5
	No	28	10.3	14.5	100.0
	Total	193	71.2	100.0	
Missing	Branching skip	43	15.9		
	Partial breakoff	28	10.3		
	No answer	7	2.6		
	Total	78	28.8		
Total		271	100.0		

Q06B Does your organization . . . Purchase and re-sell native plant materials available on the market?

		Frequency	Percent	Valid Percent	Cumulative Percent
Valid	Yes	110	40.6	59.5	59.5
	No	75	27.7	40.5	100.0
	Total	185	68.3	100.0	
Missing	Branching skip	43	15.9		
	Partial breakoff	28	10.3		
	No answer	15	5.5		
	Total	86	31.7		
Total		271	100.0		

Q07a What percentage of your organization's total native plant material sales does this represent? Native grass seed.

		Frequency	Percent	Valid Percent	Cumulative Percent
Valid	0	39	14.4	25.0	25.0
	1	5	1.8	3.2	28.2
	2	6	2.2	3.8	32.1
	3	2	.7	1.3	33.3
	5	11	4.1	7.1	40.4
	7	1	.4	.6	41.0
	8	1	.4	.6	41.7
	10	22	8.1	14.1	55.8
	12	1	.4	.6	56.4
	15	5	1.8	3.2	59.6
	18	1	.4	.6	60.3
	20	15	5.5	9.6	69.9
	24	1	.4	.6	70.5
	25	9	3.3	5.8	76.3
	30	4	1.5	2.6	78.8
	33	1	.4	.6	79.5
	35	2	.7	1.3	80.8
	40	9	3.3	5.8	86.5
	45	3	1.1	1.9	88.5
	50	8	3.0	5.1	93.6
	60	3	1.1	1.9	95.5
	75	1	.4	.6	96.2
	85	2	.7	1.3	97.4
	90	1	.4	.6	98.1
	97	1	.4	.6	98.7
	98	1	.4	.6	99.4
	100	1	.4	.6	100.0
	Total	156	57.6	100.0	
Missing	Branching skip	43	15.9		
	Partial breakoff	32	11.8		
	No answer	40	14.8		
	Total	115	42.4		
Total		271	100.0		

Q07b What percentage of your organization's total native plant material sales does this represent? Native forbs seed.

		Frequency	Percent	Valid Percent	Cumulative Percent
Valid	0	42	15.5	28.0	28.0
	1	5	1.8	3.3	31.3
	2	3	1.1	2.0	33.3
	5	4	1.5	2.7	36.0
	10	11	4.1	7.3	43.3
	15	5	1.8	3.3	46.7
	19	1	.4	.7	47.3
	20	7	2.6	4.7	52.0
	25	13	4.8	8.7	60.7
	28	1	.4	.7	61.3
	30	4	1.5	2.7	64.0
	32	1	.4	.7	64.7
	33	1	.4	.7	65.3
	35	5	1.8	3.3	68.7
	40	9	3.3	6.0	74.7
	45	7	2.6	4.7	79.3
	50	7	2.6	4.7	84.0
	55	2	.7	1.3	85.3
	60	8	3.0	5.3	90.7
	70	2	.7	1.3	92.0
	75	4	1.5	2.7	94.7
	79	1	.4	.7	95.3
	80	4	1.5	2.7	98.0
	85	2	.7	1.3	99.3
	90	1	.4	.7	100.0
	Total	150	55.4	100.0	
Missing	Branching skip	43	15.9		
	Partial breakoff	32	11.8		
	No answer	46	17.0		
	Total	121	44.6		
Total		271	100.0		

Q07c What percentage of your organization's total native plant material sales does this represent? Native shrubs seed.

		Frequency	Percent	Valid Percent	Cumulative Percent
Valid	0	25	9.2	16.3	16.3
	1	4	1.5	2.6	19.0
	2	3	1.1	2.0	20.9
	5	7	2.6	4.6	25.5
	10	14	5.2	9.2	34.6
	15	4	1.5	2.6	37.3
	18	1	.4	.7	37.9
	20	13	4.8	8.5	46.4
	22	1	.4	.7	47.1
	25	24	8.9	15.7	62.7
	30	13	4.8	8.5	71.2
	33	1	.4	.7	71.9
	34	1	.4	.7	72.5
	35	3	1.1	2.0	74.5
	38	1	.4	.7	75.2
	40	13	4.8	8.5	83.7
	45	2	.7	1.3	85.0
	50	10	3.7	6.5	91.5
	55	1	.4	.7	92.2
	60	2	.7	1.3	93.5
	70	1	.4	.7	94.1
	75	2	.7	1.3	95.4
	87	1	.4	.7	96.1
	89	1	.4	.7	96.7
	90	1	.4	.7	97.4
	100	4	1.5	2.6	100.0
	Total	153	56.5	100.0	
Missing	Branching skip	43	15.9		
	Partial breakoff	32	11.8		
	No answer	43	15.9		
	Total	118	43.5		
Total		271	100.0		

Q07d What percentage of your organization's total native plant material sales does this represent? Native tree seed.

		Frequency	Percent	Valid Percent	Cumulative Percent
Valid	0	22	8.1	14.5	14.5
	1	3	1.1	2.0	16.4
	2	3	1.1	2.0	18.4
	5	9	3.3	5.9	24.3
	8	1	.4	.7	25.0
	10	9	3.3	5.9	30.9
	13	1	.4	.7	31.6
	15	8	3.0	5.3	36.8
	17	1	.4	.7	37.5
	20	15	5.5	9.9	47.4
	22	1	.4	.7	48.0
	25	14	5.2	9.2	57.2
	30	8	3.0	5.3	62.5
	32	1	.4	.7	63.2
	33	1	.4	.7	63.8
	35	4	1.5	2.6	66.4
	40	3	1.1	2.0	68.4
	42	1	.4	.7	69.1
	45	3	1.1	2.0	71.1
	50	10	3.7	6.6	77.6
	59	1	.4	.7	78.3
	60	3	1.1	2.0	80.3
	65	1	.4	.7	80.9
	70	1	.4	.7	81.6
	80	3	1.1	2.0	83.6
	85	1	.4	.7	84.2
	89	1	.4	.7	84.9
	98	1	.4	.7	85.5
	100	22	8.1	14.5	100.0
	Total	152	56.1	100.0	
Missing	Branching skip	43	15.9		
	Partial breakoff	32	11.8		
	No answer	44	16.2		
	Total	119	43.9		
Total		271	100.0		

Q07e What percentage of your organization's total native plant material sales does this represent? Other native seed.

		Frequency	Percent	Valid Percent	Cumulative Percent
Valid	0	57	21.0	49.6	49.6
	1	1	.4	.9	50.4
	2	3	1.1	2.6	53.0
	3	3	1.1	2.6	55.7
	4	3	1.1	2.6	58.3
	5	8	3.0	7.0	65.2
	8	3	1.1	2.6	67.8
	10	11	4.1	9.6	77.4
	15	3	1.1	2.6	80.0
	20	5	1.8	4.3	84.3
	25	4	1.5	3.5	87.8
	30	3	1.1	2.6	90.4
	35	1	.4	.9	91.3
	36	1	.4	.9	92.2
	40	1	.4	.9	93.0
	50	4	1.5	3.5	96.5
	60	1	.4	.9	97.4
	74	1	.4	.9	98.3
	100	2	.7	1.7	100.0
	Total	115	42.4	100.0	
Missing	Branching skip	43	15.9		
	Partial breakoff	32	11.8		
	No answer	81	29.9		
	Total	156	57.6		
Total		271	100.0		

Q07a_dk What percentage of your organization's total native plant material sales does this represent? Check this box if you are unable to provide this information.

		Frequency	Percent	Valid Percent	Cumulative Percent
Valid	Checked	15	5.5	6.8	6.8
	Not checked	205	75.6	93.2	100.0
	Total	220	81.2	100.0	
Missing	Branching skip	43	15.9		
	Partial breakoff	8	3.0		
	Total	51	18.8		
Total		271	100.0		

Q08 Do you sell native plants or seeds sourced from different locations to match different geographical conditions or seed zones, as in different ecotypes of the same species?

		Frequency	Percent	Valid Percent	Cumulative Percent
Valid	Yes	155	57.2	69.2	69.2
	No	69	25.5	30.8	100.0
	Total	224	82.7	100.0	
Missing	Dont know (phone only)	3	1.1		
	Partial breakoff	32	11.8		
	No answer	12	4.4		
	Total	47	17.3		
Total		271	100.0		

Q09A Between 2017 and 2019 how much did this buyer type represent your annual sales of native seed and plant materials? Federal government agencies.

		Frequency	Percent	Valid Percent	Cumulative Percent
Valid	Major portion of sales	43	15.9	19.1	19.1
	Minor portion of sales	101	37.3	44.9	64.0
	None of the sales	81	29.9	36.0	100.0
	Total	225	83.0	100.0	
Missing	Partial breakoff	35	12.9		
	No answer	11	4.1		
	Total	46	17.0		
Total		271	100.0		

Q09B Between 2017 and 2019 how much did this buyer type represent your annual sales of native seed and plant materials? State government agencies.

		Frequency	Percent	Valid Percent	Cumulative Percent
Valid	Major portion of sales	62	22.9	27.4	27.4
	Minor portion of sales	115	42.4	50.9	78.3
	None of the sales	49	18.1	21.7	100.0
	Total	226	83.4	100.0	
Missing	Partial breakoff	35	12.9		
	No answer	10	3.7		
	Total	45	16.6		
Total		271	100.0		

Q09C Between 2017 and 2019 how much did this buyer type represent your annual sales of native seed and plant materials? Municipalities.

		Frequency	Percent	Valid Percent	Cumulative Percent
Valid	Major portion of sales	52	19.2	23.2	23.2
	Minor portion of sales	122	45.0	54.5	77.7
	None of the sales	50	18.5	22.3	100.0
	Total	224	82.7	100.0	
Missing	Partial breakoff	35	12.9		
	No answer	12	4.4		
	Total	47	17.3		
Total		271	100.0		

Q09D Between 2017 and 2019 how much did this buyer type represent your annual sales of native seed and plant materials? Tribal governments.

		Frequency	Percent	Valid Percent	Cumulative Percent
Valid	Major portion of sales	14	5.2	6.5	6.5
	Minor portion of sales	70	25.8	32.4	38.9
	None of the sales	132	48.7	61.1	100.0
	Total	216	79.7	100.0	
Missing	Partial breakoff	35	12.9		
	No answer	20	7.4		
	Total	55	20.3		
Total		271	100.0		

Q09E Between 2017 and 2019 how much did this buyer type represent your annual sales of native seed and plant materials? Private contractors.

		Frequency	Percent	Valid Percent	Cumulative Percent
Valid	Major portion of sales	139	51.3	60.7	60.7
	Minor portion of sales	72	26.6	31.4	92.1
	None of the sales	18	6.6	7.9	100.0
	Total	229	84.5	100.0	
Missing	Partial breakoff	35	12.9		
	No answer	7	2.6		
	Total	42	15.5		
Total		271	100.0		

Q09F Between 2017 and 2019 how much did this buyer type represent your annual sales of native seed and plant materials? Non-profit organizations.

		Frequency	Percent	Valid Percent	Cumulative Percent
Valid	Major portion of sales	50	18.5	22.1	22.1
	Minor portion of sales	129	47.6	57.1	79.2
	None of the sales	47	17.3	20.8	100.0
	Total	226	83.4	100.0	
Missing	Dont know (phone only)	1	.4		
	Partial breakoff	35	12.9		
	No answer	9	3.3		
	Total	45	16.6		
Total		271	100.0		

Q09G Between 2017 and 2019 how much did this buyer type represent your annual sales of native seed and plant materials? Farmers and ranchers.

		Frequency	Percent	Valid Percent	Cumulative Percent
Valid	Major portion of sales	44	16.2	19.6	19.6
	Minor portion of sales	108	39.9	48.0	67.6
	None of the sales	73	26.9	32.4	100.0
	Total	225	83.0	100.0	
Missing	Partial breakoff	35	12.9		
	No answer	11	4.1		
	Total	46	17.0		
Total		271	100.0		

Q10 Do you sell services?

		Frequency	Percent	Valid Percent	Cumulative Percent
Valid	Yes. Which services? Please specify:	84	31.0	36.8	36.8
	No	144	53.1	63.2	100.0
	Total	228	84.1	100.0	
Missing	Partial breakoff	36	13.3		
	No answer	7	2.6		
	Total	43	15.9		
Total		271	100.0		

Q11A Does the availability of this type of storage limit the quantity of seed you can sell? Ambient storage.

		Frequency	Percent	Valid Percent	Cumulative Percent
Valid	Limits it	35	12.9	16.7	16.7
	Does not limit it	175	64.6	83.3	100.0
	Total	210	77.5	100.0	
Missing	Dont know (phone only)	6	2.2		
	Partial breakoff	37	13.7		
	No answer	18	6.6		
	Total	61	22.5		
Total		271	100.0		

Q11B Does the availability of this type of storage limit the quantity of seed you can sell? Refrigerated storage.

		Frequency	Percent	Valid Percent	Cumulative Percent
Valid	Limits it	67	24.7	31.9	31.9
	Does not limit it	143	52.8	68.1	100.0
	Total	210	77.5	100.0	
Missing	Dont know (phone only)	7	2.6		
	Partial breakoff	37	13.7		
	No answer	17	6.3		
	Total	61	22.5		
Total		271	100.0		

Q11C Does the availability of this type of storage limit the quantity of seed you can sell? Freezer storage.

		Frequency	Percent	Valid Percent	Cumulative Percent
Valid	Limits it	61	22.5	29.8	29.8
	Does not limit it	144	53.1	70.2	100.0
	Total	205	75.6	100.0	
Missing	Dont know (phone only)	7	2.6		
	Partial breakoff	37	13.7		
	No answer	22	8.1		
	Total	66	24.4		
Total		271	100.0		

Q12A Do you use this method to communicate to potential buyers about what your business offers? Company website.

		Frequency	Percent	Valid Percent	Cumulative Percent
Valid	Use	195	72.0	85.2	85.2
	Do not use	34	12.5	14.8	100.0
	Total	229	84.5	100.0	
Missing	Partial breakoff	38	14.0		
	No answer	4	1.5		
	Total	42	15.5		
Total		271	100.0		

Q12B Do you use this method to communicate to potential buyers about what your business offers? Print catalogue.

		Frequency	Percent	Valid Percent	Cumulative Percent
Valid	Use	80	29.5	35.6	35.6
	Do not use	145	53.5	64.4	100.0
	Total	225	83.0	100.0	
Missing	Partial breakoff	38	14.0		
	No answer	8	3.0		
	Total	46	17.0		
Total		271	100.0		

Q12C Do you use this method to communicate to potential buyers about what your business offers? Vendor listings or registries.

		Frequency	Percent	Valid Percent	Cumulative Percent
Valid	Use	146	53.9	65.8	65.8
	Do not use	76	28.0	34.2	100.0
	Total	222	81.9	100.0	
Missing	Dont know (phone only)	1	.4		
	Partial breakoff	38	14.0		
	No answer	10	3.7		
	Total	49	18.1		
Total		271	100.0		

Q12D Do you use this method to communicate to potential buyers about what your business offers? Bids in response to "requests for proposals" or "seed buy requests".

		Frequency	Percent	Valid Percent	Cumulative Percent
Valid	Use	173	63.8	75.9	75.9
	Do not use	55	20.3	24.1	100.0
	Total	228	84.1	100.0	
Missing	Partial breakoff	38	14.0		
	No answer	5	1.8		
	Total	43	15.9		
Total		271	100.0		

Q12E Do you use this method to communicate to potential buyers about what your business offers? Word-of-mouth.

		Frequency	Percent	Valid Percent	Cumulative Percent
Valid	Use	219	80.8	96.5	96.5
	Do not use	8	3.0	3.5	100.0
	Total	227	83.8	100.0	
Missing	Partial breakoff	38	14.0		
	No answer	6	2.2		
	Total	44	16.2		
Total		271	100.0		

Q12F Do you use this method to communicate to potential buyers about what your business offers? Other, please specify.

		Frequency	Percent	Valid Percent	Cumulative Percent
Valid	Use	53	19.6	56.4	56.4
	Do not use	41	15.1	43.6	100.0
	Total	94	34.7	100.0	
Missing	Dont know (phone only)	1	.4		
	Partial breakoff	38	14.0		
	No answer	138	50.9		
	Total	177	65.3		
Total		271	100.0		

Q13A Do you use this type of contracting arrangement with buyers? Bids on periodic consolidated seed buys.

		Frequency	Percent	Valid Percent	Cumulative Percent
Valid	Yes	51	18.8	22.6	22.6
	No	175	64.6	77.4	100.0
	Total	226	83.4	100.0	
Missing	Dont know (phone only)	2	.7		
	Partial breakoff	38	14.0		
	No answer	5	1.8		
	Total	45	16.6		
Total		271	100.0		

Q13B Do you use this type of contracting arrangement with buyers? Other spot market sales of available seed or plant materials.

		Frequency	Percent	Valid Percent	Cumulative Percent
Valid	Yes	115	42.4	51.3	51.3
	No	109	40.2	48.7	100.0
	Total	224	82.7	100.0	
Missing	Dont know (phone only)	1	.4		
	Partial breakoff	38	14.0		
	No answer	8	3.0		
	Total	47	17.3		
Total		271	100.0		

Q13C Do you use this type of contracting arrangement with buyers? Marketing contract that specifies the type, price, quantity, and delivery date of seed or plant materials.

		Frequency	Percent	Valid Percent	Cumulative Percent
Valid	Yes	117	43.2	52.0	52.0
	No	108	39.9	48.0	100.0
	Total	225	83.0	100.0	
Missing	Partial breakoff	38	14.0		
	No answer	8	3.0		
	Total	46	17.0		
Total		271	100.0		

Q13D Do you use this type of contracting arrangement with buyers? Production contract that specifies the desired type, quantity and delivery date of seed or plant materials.

		Frequency	Percent	Valid Percent	Cumulative Percent
Valid	Yes	119	43.9	52.9	52.9
	No	106	39.1	47.1	100.0
	Total	225	83.0	100.0	
Missing	Dont know (phone only)	1	.4		
	Partial breakoff	38	14.0		
	No answer	7	2.6		
	Total	46	17.0		
Total		271	100.0		

Q14A How important is this characteristic when considering a contract? Price guarantee.

		Frequency	Percent	Valid Percent	Cumulative Percent
Valid	Very important	125	46.1	55.6	55.6
	Somewhat important	49	18.1	21.8	77.3
	Not important	10	3.7	4.4	81.8
	No opinion	41	15.1	18.2	100.0
	Total	225	83.0	100.0	
Missing	Partial breakoff	40	14.8		
	No answer	6	2.2		
	Total	46	17.0		
Total		271	100.0		

Q14B How important is this characteristic when considering a contract? Purchase guarantee for predetermined quantity.

		Frequency	Percent	Valid Percent	Cumulative Percent
Valid	Very important	124	45.8	55.4	55.4
	Somewhat important	52	19.2	23.2	78.6
	Not important	10	3.7	4.5	83.0
	No opinion	38	14.0	17.0	100.0
	Total	224	82.7	100.0	
Missing	Partial breakoff	40	14.8		
	No answer	7	2.6		
	Total	47	17.3		
Total		271	100.0		

Q14C How important is this characteristic when considering a contract? Production cost sharing.

		Frequency	Percent	Valid Percent	Cumulative Percent
Valid	Very important	20	7.4	9.3	9.3
	Somewhat important	43	15.9	20.0	29.3
	Not important	78	28.8	36.3	65.6
	No opinion	74	27.3	34.4	100.0
	Total	215	79.3	100.0	
Missing	Dont know (phone only)	2	.7		
	Partial breakoff	40	14.8		
	No answer	14	5.2		
	Total	56	20.7		
Total		271	100.0		

Q14D How important is this characteristic when considering a contract? Multi-production cycle contract.

		Frequency	Percent	Valid Percent	Cumulative Percent
Valid	Very important	24	8.9	10.9	10.9
	Somewhat important	52	19.2	23.6	34.5
	Not important	68	25.1	30.9	65.5
	No opinion	76	28.0	34.5	100.0
	Total	220	81.2	100.0	
Missing	Dont know (phone only)	3	1.1		
	Partial breakoff	40	14.8		
	No answer	8	3.0		
	Total	51	18.8		
Total		271	100.0		

Q14E How important is this characteristic when considering a contract? Delivery timeline.

		Frequency	Percent	Valid Percent	Cumulative Percent
Valid	Very important	122	45.0	55.0	55.0
	Somewhat important	60	22.1	27.0	82.0
	Not important	10	3.7	4.5	86.5
	No opinion	30	11.1	13.5	100.0
	Total	222	81.9	100.0	
Missing	Dont know (phone only)	1	.4		
	Partial breakoff	40	14.8		
	No answer	8	3.0		
	Total	49	18.1		
Total		271	100.0		

Q14F How important is this characteristic when considering a contract? Prior positive experience with buyer.

		Frequency	Percent	Valid Percent	Cumulative Percent
Valid	Very important	124	45.8	55.4	55.4
	Somewhat important	68	25.1	30.4	85.7
	Not important	8	3.0	3.6	89.3
	No opinion	24	8.9	10.7	100.0
	Total	224	82.7	100.0	
Missing	Partial breakoff	40	14.8		
	No answer	7	2.6		
	Total	47	17.3		
Total		271	100.0		

Q15 Relative to when you would start to produce native seeds or plant materials for sale, when do you normally sign a contract to sell these seeds or plant materials?

		Frequency	Percent	Valid Percent	Cumulative Percent
Valid	Prior to when you would collect or purchase foundation seed	29	10.7	14.5	14.5
	Prior to when you would begin multiplying from foundation seed in your possession	8	3.0	4.0	18.5
	Prior to when you would begin producing the seeds or plant materials for sale	44	16.2	22.0	40.5
	After seed production begins but before harvest of all seed or plant material to be sold	25	9.2	12.5	53.0
	After seed production is complete	94	34.7	47.0	100.0
	Total	200	73.8	100.0	
Missing	Dont know (phone only)	5	1.8		
	Partial breakoff	42	15.5		
	No answer	24	8.9		
	Total	71	26.2		
Total		271	100.0		

Q16A What role does this type of information play in how you anticipate future demand for native seed or plant materials in order to plan ahead? BLM seed buy history.

		Frequency	Percent	Valid Percent	Cumulative Percent
Valid	Major role	14	5.2	6.5	6.5
	Moderate role	14	5.2	6.5	13.0
	Minor role	29	10.7	13.5	26.5
	No role	158	58.3	73.5	100.0
	Total	215	79.3	100.0	
Missing	Dont know (phone only)	2	.7		
	Partial breakoff	43	15.9		
	No answer	11	4.1		
	Total	56	20.7		
Total		271	100.0		

Q16B What role does this type of information play in how you anticipate future demand for native seed or plant materials in order to plan ahead? Past purchases or requests from other buyers.

		Frequency	Percent	Valid Percent	Cumulative Percent
Valid	Major role	122	45.0	55.5	55.5
	Moderate role	62	22.9	28.2	83.6
	Minor role	17	6.3	7.7	91.4
	No role	19	7.0	8.6	100.0
	Total	220	81.2	100.0	
Missing	Partial breakoff	43	15.9		
	No answer	8	3.0		
	Total	51	18.8		
Total		271	100.0		

Q16C What role does this type of information play in how you anticipate future demand for native seed or plant materials in order to plan ahead? Information about forward contracting opportunities.

		Frequency	Percent	Valid Percent	Cumulative Percent
Valid	Major role	35	12.9	16.4	16.4
	Moderate role	59	21.8	27.7	44.1
	Minor role	46	17.0	21.6	65.7
	No role	73	26.9	34.3	100.0
	Total	213	78.6	100.0	
Missing	Dont know (phone only)	2	.7		
	Partial breakoff	43	15.9		
	No answer	13	4.8		
	Total	58	21.4		
Total		271	100.0		

Q16D What role does this type of information play in how you anticipate future demand for native seed or plant materials in order to plan ahead? Conservation Research Program funding and plans.

		Frequency	Percent	Valid Percent	Cumulative Percent
Valid	Major role	24	8.9	11.2	11.2
	Moderate role	37	13.7	17.3	28.5
	Minor role	62	22.9	29.0	57.5
	No role	91	33.6	42.5	100.0
	Total	214	79.0	100.0	
Missing	Dont know (phone only)	2	.7		
	Partial breakoff	43	15.9		
	No answer	12	4.4		
	Total	57	21.0		
Total		271	100.0		

Q16E What role does this type of information play in how you anticipate future demand for native seed or plant materials in order to plan ahead? Wildfire activity.

		Frequency	Percent	Valid Percent	Cumulative Percent
Valid	Major role	11	4.1	5.1	5.1
	Moderate role	29	10.7	13.4	18.4
	Minor role	58	21.4	26.7	45.2
	No role	119	43.9	54.8	100.0
	Total	217	80.1	100.0	
Missing	Partial breakoff	43	15.9		
	No answer	11	4.1		
	Total	54	19.9		
Total		271	100.0		

Q16F What role does this type of information play in how you anticipate future demand for native seed or plant materials in order to plan ahead? Oil and gas leases.

		Frequency	Percent	Valid Percent	Cumulative Percent
Valid	Major role	4	1.5	1.9	1.9
	Moderate role	9	3.3	4.2	6.1
	Minor role	27	10.0	12.6	18.7
	No role	174	64.2	81.3	100.0
	Total	214	79.0	100.0	
Missing	Partial breakoff	43	15.9		
	No answer	14	5.2		
	Total	57	21.0		
Total		271	100.0		

Q16G What role does this type of information play in how you anticipate future demand for native seed or plant materials in order to plan ahead? Mining operations.

		Frequency	Percent	Valid Percent	Cumulative Percent
Valid	Major role	6	2.2	2.8	2.8
	Moderate role	15	5.5	7.0	9.9
	Minor role	33	12.2	15.5	25.4
	No role	159	58.7	74.6	100.0
	Total	213	78.6	100.0	
Missing	Partial breakoff	43	15.9		
	No answer	15	5.5		
	Total	58	21.4		
Total		271	100.0		

Q16H What role does this type of information play in how you anticipate future demand for native seed or plant materials in order to plan ahead? Native seed use trends.

		Frequency	Percent	Valid Percent	Cumulative Percent
Valid	Major role	26	9.6	12.1	12.1
	Moderate role	50	18.5	23.3	35.3
	Minor role	50	18.5	23.3	58.6
	No role	89	32.8	41.4	100.0
	Total	215	79.3	100.0	
Missing	Partial breakoff	43	15.9		
	No answer	13	4.8		
	Total	56	20.7		
Total		271	100.0		

Q16I What role does this type of information play in how you anticipate future demand for native seed or plant materials in order to plan ahead? Demand for urban and home landscaping.

		Frequency	Percent	Valid Percent	Cumulative Percent
Valid	Major role	70	25.8	32.1	32.1
	Moderate role	56	20.7	25.7	57.8
	Minor role	43	15.9	19.7	77.5
	No role	49	18.1	22.5	100.0
	Total	218	80.4	100.0	
Missing	Dont know (phone only)	1	.4		
	Partial breakoff	43	15.9		
	No answer	9	3.3		
	Total	53	19.6		
Total		271	100.0		

Q16J What role does this type of information play in how you anticipate future demand for native seed or plant materials in order to plan ahead? Other, please describe.

		Frequency	Percent	Valid Percent	Cumulative Percent
Valid	Major role	28	10.3	44.4	44.4
	Moderate role	10	3.7	15.9	60.3
	Minor role	1	.4	1.6	61.9
	No role	24	8.9	38.1	100.0
	Total	63	23.2	100.0	
Missing	-7	1	.4		
	Partial breakoff	43	15.9		
	No answer	164	60.5		
	Total	208	76.8		
Total		271	100.0		

Q17 Between 2017 and 2019, approximately what percentage of your inventory of native seed and plant materials was left unsold at the end of the marketing year?

		Frequency	Percent	Valid Percent	Cumulative Percent
Valid	0	16	5.9	7.5	7.5
	1	4	1.5	1.9	9.3
	2	4	1.5	1.9	11.2
	3	3	1.1	1.4	12.6
	5	14	5.2	6.5	19.2
	7	1	.4	.5	19.6
	8	1	.4	.5	20.1
	9	1	.4	.5	20.6
	10	24	8.9	11.2	31.8
	12	1	.4	.5	32.2
	15	18	6.6	8.4	40.7
	17	1	.4	.5	41.1
	20	29	10.7	13.6	54.7
	25	17	6.3	7.9	62.6
	30	29	10.7	13.6	76.2
	33	1	.4	.5	76.6
	35	9	3.3	4.2	80.8
	40	12	4.4	5.6	86.4
	45	1	.4	.5	86.9
	50	19	7.0	8.9	95.8
	60	4	1.5	1.9	97.7
	70	2	.7	.9	98.6
	75	2	.7	.9	99.5
	90	1	.4	.5	100.0
	Total	214	79.0	100.0	
Missing	Partial breakoff	43	15.9		
	No answer	14	5.2		
	Total	57	21.0		
Total		271	100.0		

Q18 Between 2017 and 2019, was the average percentage of native seed and plant materials that was left unsold at the end of the marketing year . . .

		Frequency	Percent	Valid Percent	Cumulative Percent
Valid	Less than you anticipated	28	10.3	13.0	13.0
	About what you anticipated	166	61.3	77.2	90.2
	More than you anticipated	21	7.7	9.8	100.0
	Total	215	79.3	100.0	
Missing	Dont know (phone only)	1	.4		
	Partial breakoff	43	15.9		
	No answer	12	4.4		
	Total	56	20.7		
Total		271	100.0		

Q19A How much of a challenge is this for you as a supplier of native seed or plant materials? Insufficient demand.

		Frequency	Percent	Valid Percent	Cumulative Percent
Valid	Major challenge	33	12.2	15.6	15.6
	Moderate challenge	64	23.6	30.3	46.0
	Minor challenge	56	20.7	26.5	72.5
	Not a challenge	58	21.4	27.5	100.0
	Total	211	77.9	100.0	
Missing	Partial breakoff	44	16.2		
	No answer	16	5.9		
	Total	60	22.1		
Total		271	100.0		

Q19B How much of a challenge is this for you as a supplier of native seed or plant materials? Unpredictable demand.

		Frequency	Percent	Valid Percent	Cumulative Percent
Valid	Major challenge	89	32.8	41.2	41.2
	Moderate challenge	72	26.6	33.3	74.5
	Minor challenge	39	14.4	18.1	92.6
	Not a challenge	16	5.9	7.4	100.0
	Total	216	79.7	100.0	
Missing	Dont know (phone only)	1	.4		
	Partial breakoff	44	16.2		
	No answer	10	3.7		
	Total	55	20.3		
Total		271	100.0		

Q19C How much of a challenge is this for you as a supplier of native seed or plant materials? Demand for plants without propagation protocols.

		Frequency	Percent	Valid Percent	Cumulative Percent
Valid	Major challenge	22	8.1	10.8	10.8
	Moderate challenge	26	9.6	12.7	23.5
	Minor challenge	68	25.1	33.3	56.9
	Not a challenge	88	32.5	43.1	100.0
	Total	204	75.3	100.0	
Missing	Dont know (phone only)	5	1.8		
	Partial breakoff	44	16.2		
	No answer	18	6.6		
	Total	67	24.7		
Total		271	100.0		

Q19D How much of a challenge is this for you as a supplier of native seed or plant materials? Lack of seed testing protocols.

		Frequency	Percent	Valid Percent	Cumulative Percent
Valid	Major challenge	31	11.4	15.0	15.0
	Moderate challenge	40	14.8	19.3	34.3
	Minor challenge	59	21.8	28.5	62.8
	Not a challenge	77	28.4	37.2	100.0
	Total	207	76.4	100.0	
Missing	Dont know (phone only)	3	1.1		
	Partial breakoff	44	16.2		
	No answer	17	6.3		
	Total	64	23.6		
Total		271	100.0		

Q19E How much of a challenge is this for you as a supplier of native seed or plant materials? Species difficult to grow.

		Frequency	Percent	Valid Percent	Cumulative Percent
Valid	Major challenge	45	16.6	21.2	21.2
	Moderate challenge	82	30.3	38.7	59.9
	Minor challenge	62	22.9	29.2	89.2
	Not a challenge	23	8.5	10.8	100.0
	Total	212	78.2	100.0	
Missing	Partial breakoff	44	16.2		
	No answer	15	5.5		
	Total	59	21.8		
Total		271	100.0		

Q19F How much of a challenge is this for you as a supplier of native seed or plant materials? Lack of stock seed from appropriate seed zones or other specified locations.

		Frequency	Percent	Valid Percent	Cumulative Percent
Valid	Major challenge	54	19.9	25.4	25.4
	Moderate challenge	51	18.8	23.9	49.3
	Minor challenge	57	21.0	26.8	76.1
	Not a challenge	51	18.8	23.9	100.0
	Total	213	78.6	100.0	
Missing	Partial breakoff	44	16.2		
	No answer	14	5.2		
	Total	58	21.4		
Total		271	100.0		

Q19G How much of a challenge is this for you as a supplier of native seed or plant materials? Other, please describe.

		Frequency	Percent	Valid Percent	Cumulative Percent
Valid	Major challenge	23	8.5	35.9	35.9
	Moderate challenge	12	4.4	18.8	54.7
	Minor challenge	4	1.5	6.3	60.9
	Not a challenge	25	9.2	39.1	100.0
	Total	64	23.6	100.0	
Missing	Dont know (phone only)	2	.7		
	Partial breakoff	44	16.2		
	No answer	161	59.4		
	Total	207	76.4		
Total		271	100.0		

Q20_O What suggestions do you have that might help address the challenges you have encountered as a supplier of native seeds or plant materials? (Open-ended response coding frequencies)

CODE	count
Planning	31
Coordination	20
Awareness, education	25
Source	15
Funding	14
Policy	10
Value shift	9
Regional focus	7
Seed specific	8
Markets or economics	10
Standards	7
Labor	5
Access	4
Other	2
Not in Business	2
None, No comment, no suggestions, etc.	6

Q21 If you were to anticipate higher demand for directly wild-collected native seed, would your business be able to expand to collect more seed?

		Frequency	Percent	Valid Percent	Cumulative Percent
Valid	Likely yes	69	25.5	61.6	61.6
	Likely no	43	15.9	38.4	100.0
	Total	112	41.3	100.0	
Missing	Refusal (phone only)	1	.4		
	Dont know (phone only)	1	.4		
	Branching skip	102	37.6		
	Partial breakoff	48	17.7		
	No answer	7	2.6		
	Total	159	58.7		
Total		271	100.0		

Q22 To the extent you are aware, are there major barriers or disincentives to wild-collecting native seed?

		Frequency	Percent	Valid Percent	Cumulative Percent
Valid	Yes. What are those barriers?	143	52.8	70.8	70.8
	No	59	21.8	29.2	100.0
	Total	202	74.5	100.0	
Missing	Dont know (phone only)	2	.7		
	Partial breakoff	48	17.7		
	No answer	19	7.0		
	Total	69	25.5		
Total		271	100.0		

Q22_O To the extent you are aware, are there major barriers or disincentives to wild-collecting native seed? What are those barriers? (Open-ended response coding frequencies)

CODE	count
Access to land	62
Availability of seed/land	26
Labor	22
Depletion	17
Cost	13
Time	12
Markets	10
Knowledge	10
Environmental factors	7
Ethical reasons	6
Quality of seed	6
Resources Available	4
Other	2

Q23 If you were to anticipate higher demand, would your business be able to expand to grow more native plants with the goal of producing and selling native seed?

		Frequency	Percent	Valid Percent	Cumulative Percent
Valid	Likely yes	77	28.4	75.5	75.5
	Likely no	25	9.2	24.5	100.0
	Total	102	37.6	100.0	
Missing	Dont know (phone only)	2	.7		
	Branching skip	118	43.5		
	Partial breakoff	48	17.7		
	No answer	1	.4		
	Total	169	62.4		
Total		271	100.0		

Q24 To the extent you are aware, are there any major barriers or disincentives to growing native plants with the goal of producing and selling native seed?

		Frequency	Percent	Valid Percent	Cumulative Percent
Valid	Yes. What are those barriers?	83	30.6	42.8	42.8
	No	111	41.0	57.2	100.0
	Total	194	71.6	100.0	
Missing	Dont know (phone only)	2	.7		
	Partial breakoff	48	17.7		
	No answer	27	10.0		
	Total	77	28.4		
Total		271	100.0		

Q24_O To the extent you are aware, are there any major barriers or disincentives to growing native plants with the goal of producing and selling native seed? What are those barriers? (Open-ended response coding frequencies)

CODE	count
Markets	24
Seed/plant viability	14
Cost	14
Availability	8
Space	8
Environmental factors	7
Labor	6
Knowledge	6
Time	5
Resources	4
Zoning Requirements	3
Permits	3
Other	3

Q25 If you were to anticipate higher demand, would your business be able to expand to grow and sell more plant materials?

		Frequency	Percent	Valid Percent	Cumulative Percent
Valid	Likely yes	124	45.8	83.2	83.2
	Likely no	25	9.2	16.8	100.0
	Total	149	55.0	100.0	
Missing	Dont know (phone only)	1	.4		
	Branching skip	71	26.2		
	Partial breakoff	48	17.7		
	No answer	2	.7		
	Total	122	45.0		
Total		271	100.0		

Q26 To the extent you are aware, are there any major barriers or disincentives to growing and selling plant materials?

		Frequency	Percent	Valid Percent	Cumulative Percent
Valid	Yes. What are those barriers?	91	33.6	45.3	45.3
	No	110	40.6	54.7	100.0
	Total	201	74.2	100.0	
Missing	Dont know (phone only)	1	.4		
	Partial breakoff	48	17.7		
	No answer	21	7.7		
	Total	70	25.8		
Total		271	100.0		

Q26_O To the extent you are aware, are there any major barriers or disincentives to growing and selling plant materials? What are those barriers? (Open-ended response coding frequencies)

CODE	count
Markets	29
Knowledge, education, perception	13
Viability	9
Cost	10
Labor	8
Availability	8
Time	3
Space	3
Environment	4
Permits	1
See previous comment	5
Other	6

Q27_O Do you have any other suggestions that could improve how the native seed market functions? (Open-ended response coding frequencies)

CODE	count
Coordination	21
Education	14
Standards	12
Regional focus	11
Predictability	9
Policy	8
Availability	5
Transparency	4
Market/Economy	4
Resilience	2
Specialty crop focus	2
More Collectors or producers	2
Monopolies	1
Atypical Grower	1
Research	1
None	14
Other	8

Q28 Between 2017 and 2019, which of the following best describes your business's average annual sales and operating revenues?

		Frequency	Percent	Valid Percent	Cumulative Percent
Valid	Less than $100,000	49	18.1	23.4	23.4
	$100,000-$499,999	63	23.2	30.1	53.6
	$500,000-$999,999	26	9.6	12.4	66.0
	$1,000,000-$4,999,999	41	15.1	19.6	85.6
	$5,000,000-$10,000,000	20	7.4	9.6	95.2
	Over $10,000,000	10	3.7	4.8	100.0
	Total	209	77.1	100.0	
Missing	Refusal (phone only)	1	.4		
	Dont know (phone only)	1	.4		
	Partial breakoff	48	17.7		
	No answer	12	4.4		
	Total	62	22.9		
Total		271	100.0		

Q29 Roughly what percentage of your average annual sales and operating revenues are from the sale of native seed and plant materials?

		Frequency	Percent	Valid Percent	Cumulative Percent
Valid	1	4	1.5	1.9	1.9
	3	1	.4	.5	2.4
	5	6	2.2	2.8	5.2
	6	1	.4	.5	5.7
	8	2	.7	.9	6.6
	10	4	1.5	1.9	8.5
	15	8	3.0	3.8	12.3
	20	10	3.7	4.7	17.1
	25	3	1.1	1.4	18.5
	27	1	.4	.5	19.0
	30	8	3.0	3.8	22.7
	33	2	.7	.9	23.7
	35	1	.4	.5	24.2
	40	3	1.1	1.4	25.6
	45	2	.7	.9	26.5
	50	13	4.8	6.2	32.7
	55	1	.4	.5	33.2
	60	11	4.1	5.2	38.4
	65	5	1.8	2.4	40.8
	70	8	3.0	3.8	44.5
	75	7	2.6	3.3	47.9
	80	10	3.7	4.7	52.6
	85	7	2.6	3.3	55.9
	90	15	5.5	7.1	63.0
	95	8	3.0	3.8	66.8
	98	4	1.5	1.9	68.7
	99	7	2.6	3.3	72.0
	100	59	21.8	28.0	100.0
	Total	211	77.9	100.0	
Missing	Partial breakoff	47	17.3		
	No answer	13	4.8		
	Total	60	22.1		
Total		271	100.0		

Appendix 2H

Public Information-Gathering Sessions
In-Person and Virtual Agendas

COMMITTEE ON AN ASSESSMENT OF NATIVE SEED NEEDS AND CAPACITIES

AUGUST 18-19, 2019
National Academy of Sciences Lecture Room
Washington, DC

Sunday, August 18, 2019
OPEN SESSION – Public Welcome (Zoom and in-person)

3:00 pm	Welcome and Introductions
3:15 pm	Discussion with Sponsor
	Peggy Olwell, *Plant Conservation Program Lead,* Bureau of Land Management (BLM), Department of the Interior (in person)
4:30 pm	Roles of different federal agencies to the native seed system: USFWS
	Sarah Kulpa, *Botanist,* US Fish and Wildlife Service (USFWS) (Nevada), Department of the Interior (via Zoom)
5:15 pm	Public comment period
5:30 pm	Adjourn

Monday, August 19, 2019
OPEN SESSION – Public Welcome (Zoom and in-person)

10:00 am	Opening remarks – Susan Harrison, *Committee Chairwoman*
10:10 am	Roles of different federal agencies to the native seed system: USFS
	William Carromero, *National Botanist,* US Forest Service (USFS) (in person) and Carol Spurrier, *Ecologist,* USFS (by phone)
10:50 am	Roles of different federal agencies to the native seed system: USDA/NRCS
	John Englert, *National Program Leader—Plant Materials,* Natural Resource Conservation Service (NRCS), US Department of Agriculture (in person)

11:30 am	Seed company perspectives on the supply and demand for native seed
	Dustin Terrell, *Partner,* Buffalo Brand Seed, LLC (via zoom)
12:10 pm	Brief break
12:40 pm	Seed company perspectives on the supply and demand for native seed
	Ed Kleiner, *General Manager,* Comstock Seed, LLC (via zoom)
1:20 pm	Roles of different federal agencies to the native seed system: USDA/NIFA
	Jim Dobrowolski, *National Program Leader,* National Institute for Food and Agriculture, US Department of Agriculture (in person)
1:50 pm	Public comment period
2:05 pm	Adjourn public session

OCTOBER 10, 2019
USFS Facility (Sage and Lupine Rooms)
Bend, Oregon

Thursday, October 10, 2019
OPEN SESSION — Public Welcome (Zoom and in-person)

8:00 am	Holly Jewkes, *Supervisor,* Deschutes National Forest, USFS — Welcome
8:15 am	Susan Harrison, *Committee Chair* — Appreciation and Introductions
8:25 am	Robin Schoen, *Study Director* — Overview of the Assessment Process
8:40 am	Pause in Zoom broadcast
8:45 am	Visit to the Bend Seed Extractory — Kayla Herriman, *USFS Extractory Manager*
10:45 am	Return to Conference Room, Prepare for Open Zoom Session
11:00 am	Susan Harrison, *Committee Chair* — Welcome, Review of the Agenda, and Introductions
11:20 am	Vicky Erickson, *Geneticist,* Pacific Northwest Region, US Forest Service
12:15 pm	Barry Schrumpf, *Seed Certification Specialist,* Oregon State University Seed Certification Service/ Oregon Seed Association
1:00 pm	Lunch break — pause in Zoom broadcast
1:45 pm	Trish Roller and Ricardo Galvan, *BLM National Seed Warehouse System, seed buys and contracting considerations* (via Zoom)
2:35 pm	Greg Eckert, *Restoration Ecologist,* Biological Resources Division, National Park Service (via Zoom)
3:30 pm	Break — pause in Zoom broadcast
3:45 pm	Jerry Benson, *President,* BFI Native Seeds
4:40 pm	Bonnie Harper-Lore, *Restoration Ecologist* (retired) Federal Highway Administration Department of Transportation (via Zoom)
5:30 pm	Public comments
5:45 pm	Adjournment of Open Session

JANUARY 28-29, 2020
National Academy of Sciences Beckman Center
Irvine, California

Tuesday, January 28, 2020
OPEN SESSION — Public Welcome (Zoom and in-person)

12:40 pm	Welcome, Susan Harrison
12:45 pm	US Tribal Nursery Council and Understanding Tribal Needs for Native Seed
	Jeremy Pinto, *US Forest Service* (Moscow, Idaho)

1:30 pm   Economics of the native seed supply
                 Charles Perring, *Arizona State University*
2:45 pm   End of Open Session

Wednesday, January 29, 2020
OPEN SESSION — Public Welcome (Zoom and in-person)
9:00 am   Welcome, Susan Harrison
                 The Role of Native Seeds on Highway Roadsides
                 John Krouse, *Maryland Department of Transportation*
                 Ken Greave, *Minnesota Department of Transportation*
10:00 am   End of Open Session

FEBRUARY 22, 2021
Via Zoom

Monday, February 22, 2021
OPEN SESSION — Public Welcome (via Zoom)
3:00 pm   Welcome, Brief Committee Introductions and Introduction of Panelists
                 Susan Harrison, *Committee Chair*
3:10 pm   Panel discussion, The Nature Conservancy Representatives
                 Olga Kildisheva, *Innovative Restoration Project Manager,* Sagebrush Sea Program, TNC Oregon
                 Catherine Schloegel, *Watershed Forest Manager,* TNC Colorado
                 Anne Bradley, *Forest Conservation Program Director (or Collin Haffey, Forest and Watershed
                 Health Manager)*, TNC New Mexico
                 Kevin Badik, *Rangeland Ecologist,* TNC Nevada
                 — Description of TNC projects and responses to key questions
                 — Questions from the committee
4:15 pm   Adjourn public session

MARCH 8, 2021
Via Zoom

Monday, March 8, 2021
OPEN SESSION — Public Welcome (via Zoom)
4:00 pm   Welcome, Brief Committee Introductions and Introduction of Panelists
                 Susan Harrison, *Committee Chair*
4:10 pm   Panel Discussants 1:
                 Rick Iovanna, *Agricultural Economist,* USDA Natural Resources Conservation Service,
                 Washington, DC
                 Bryan Pratt, *Research Agricultural Economist,* USDA Economic Research Service, Kansas City,
                 MO
4:30 pm   Clarifying Q&A for Panel 1 from Committee and other Discussants
4:40 pm   Panel Discussants 2:
                 Laura Jackson, *Director and Professor of Biology,* Tallgrass Prairie Center, University of Northern
                 Iowa, Cedar Falls, IA
                 Justin Meissen, *Research and Restoration Program,* Tallgrass Prairie Center, University of
                 Northern Iowa

Laura Walter, *Plant Materials Program Manager,* Tallgrass Prairie Center of the University of Northern Iowa

Kristine Nemec, *Integrated Roadside Vegetation Manager,* Tallgrass Prairie Center of the University of Northern Iowa

5:00 pm	Clarifying Q&A for Panel 2 from Committee and other Discussants
5:10 pm	Committee Q&A and Discussion with all Discussants
5:30 pm	Adjourn

MARCH 16, 2021
Via Zoom

Tuesday, March 16, 2021

OPEN SESSION — Public Welcome (via Zoom)

4:00 pm	Welcome, Brief Committee Introductions and Introduction of Discussants
	Susan Harrison, *Committee Chair*
4:05 pm	Monica Pokorny, *Plant Materials Specialist,* USDA Natural Resources Conservation Service, Washington, DC
	Gord Pearse, *Agronomist,* Bruce Seed Farm, and *Past-President,* Montana Seed Trade Association
4:35 pm	Committee Q&A
5:00 pm	Adjourn Public Session

MARCH 31, 2021
Via Zoom

Wednesday, March 31, 2021

OPEN SESSION — Public Welcome (via Zoom)

1:00 pm	Welcome, Brief Committee Introductions and Introduction of John Englert
	Susan Harrison, *Committee Chair*
1:05 pm	John Englert, *National Program Leader,* USDA Natural Resources Conservation Service
1:35 pm	Committee Q&A
2:00 pm	Adjourn Public Session

APRIL 12, 2021
Via Zoom

Monday, April 12, 2021

OPEN SESSION — Public Welcome (via Zoom)

4:00 pm	Welcome, Brief Committee Introductions and Introduction of Presenters
	Susan Harrison, *Committee Chair*
4:05 pm	Keith Pawelek, *Associate Director,* Texas Native Seeds
4:30 pm	Committee Q&A
4:45 pm	Ed Toth, *Founder and Director,* Greenbelt Native Plants Center, NYC Parks
5:10 pm	Committee Q&A
5:25 pm	Additional questions, discussion if needed
5:30 pm	Adjourn

APRIL 20, 2021
Via Zoom

Tuesday, April 20, 2021

OPEN SESSION — Public Welcome (via Zoom)

4:00 pm	Welcome, Brief Committee Introductions and Introduction of Ed Toth
	Susan Harrison, *Committee Chair*
4:10 pm	Ed Toth, *Founder and Director,* Greenbelt Native Plants Center, NYC Parks
4:45 pm	Committee Q&A
	Questions from public (submitted in chat)
5:30 pm	Adjourn

MAY 4, 2021
Via Zoom

Tuesday, May 4, 2021

OPEN SESSION — Public Welcome (via Zoom)

3:30 pm	Susan Harrison, *Committee Chairperson*
	Welcome and Introductions
3:35 pm	Gil Waibel, *Seed Physiologist Technologist Lead,* Bayer CropSciences
	Presentation on Seed Testing
4:15 pm	Committee Q&A
4:30 pm	Dustin Terrell, *Buffalo Brand Seeds*
	Andy Ernst, *Ernst Conservation Seeds*
	Comments on Testing from the Seed Supplier perspective
4:45 pm	Committee Q&A
5:00 pm	Adjourn

JUNE 29, 2021
Via Zoom

Tuesday, June 29, 2021

OPEN SESSION — Public Welcome (via Zoom)

3:30 pm	Susan Harrison, *Committee Chair* — Welcome
3:40 pm	Molly Anthony, *Program Lead,* BLM Emergency Stabilization and Rehabilitation Program
	BLM Post-Wildfire Recovery: Emergency Stabilization & Burned Area Rehabilitation
4:00 pm	Q&A (Committee)
4:20 pm	Anne S. Halford, *State Botanist,* BLM Idaho [Pre-Taped]
	Native Seed Production Tools to Procure Increase and Distribute Source Identified Seed via Seed Transfer Zones at an Ecoregional Scale in the Western U.S
4:40 pm	Patricia Roller, *National Seed Coordinator,* BLM National Seed Warehouse System and Ricardo Galvan, *Administrative & Financial Seed Specialist,* BLM National Seed Warehouse System
	Trends in the Periodic Regional Seed Purchases of the BLM
5:10 pm	Q&A (Committee)
5:30 pm	Adjourn

NOVEMBER 12, 2021
Via Zoom

Friday, November 12, 2021

OPEN SESSION — Public Information-Gathering Session

11:00 am	Susan Harrison (*Chair*) — Welcome and disclaimer
11:05 am	Vera Smith and Bart Johansen-Harris, of *Defenders of Wildlife*
11:25 am	Q&A from committee
11:35 am	Josh Osher and Laura Welp, of *Western Watersheds*
11:55 am	Q&A from c ommittee
12:05 pm	Cristina Eisenberg, *Oregon State University*
12:25 pm	Q&A from committee
12:35 pm	Discussion/Q&A from committee